THE CARBON AGE

THE CARBON AGE

*How Life's Core Element Has
Become Civilization's Greatest Threat*

ERIC ROSTON

Walker & Company
New York

Published by Walker Publishing Company, Inc., New York
Distributed to the trade by Macmillan

Art Credits
Page 21: T. A. Rector (NOAO/AURA/NSF) and Hubble Heritage Team (STScI/AURA/NASA). Page 32 reproduced with permission from John van Wyhe ed., *The Complete Work of Charles Darwin Online* (http://darwin-online.org.uk). Page 70 © Markus Geisen/Natural History Museum, London, used with permission. Page 84 from *The Evolution of Man: A Popular Exposition of the Principal Points of Human Ontogeny and Philogeny*, by Ernst Haeckel. Vol. 2. New York: D. Appleton and Company, 1897. Page 91 courtesy Scott Wing. Page 113: Washington, D.C. Public Library, Washingtoniana Division, Collection 57. Page 126 copyright Smithsonian Institution, NMAH/Transportation, used with permission. Page 158: Edgar Fahs Smith Collection, University of Pennsylvania Library. Pages 174 and 176: Scripps Institution of Oceanography, UC San Diego. Page 218 © 2001 Cédric Thévenet, used with permission (http://creativecommons.org/licenses/by-sa/3.0). Page 223 released under the GNU Free Documentation License.

LIBRARY OF CONGRESS CATALOGING-IN-PUBLICATION DATA

Roston, Eric.
    The carbon age : how life's core element has become civilization's greatest threat / Eric Roston.—1st U.S. ed.
        p. cm.
    Includes bibliographical references.
    ISBN-13: 978-0-8027-1557-9 (alk. paper)
    ISBN 10: 0-8027-1557-5
    1. Carbon cycle (Biogeochemistry)   I. Title.
QH344.R67 2008
577'.144—dc22
2008002754

Visit Walker & Company's Web site at www.walkerbooks.com

First U.S. edition 2008

10 9 8 7 6 5 4 3 2 1

Typeset by Westchester Book Group
Printed in the United States of America by Quebecor World Fairfield

For Karen, my life's core element

CONTENTS

Author's Note

Those of us who greet the morning air with an instinctive temperature estimate in Fahrenheit (F) have no visceral understanding of Celsius (C). But the scientific community uses only the Celsius scale, and hewing close to Celsius in science actually eases communication and avoids confusion. So, where I talk about temperatures, it will be noted first in Celsius, then if necessary, in Fahrenheit.

The Earth's average temperature is about 14°C, or 57°F. In the twentieth century, the planet warmed about 0.6°C or 1°F. Many scientists talk about an upper limit of allowable warming as 2°C, or 3.6°F. Celsius is used to estimate temperatures in the deep past, too. For example, according to one model, volcanism gone wild over a million years caused a 6°C temperature rise 251 million years ago that killed off 95 percent of species on Earth.[1]

PROLOGUE
The Strange Case of Carbon

You will be astonished when I tell you what this curious play of carbon amounts to.

—Michael Faraday

Welcome to the Carbon Age.

These days we hear the word "carbon" in the headlines frequently, but we rarely understand much about what it is. Carbon is a culprit in creating one crisis after the next. Carbon dioxide emissions warm the climate. The volatile Middle East explodes atop its stores of volatile hydrocarbons, otherwise known as oil. Carbohydrates' popularity waxes and wanes with diet crazes. Pharmaceuticals, which often rely on oil-based feedstock, command ever-higher prices. The U.S. military demands carbon fiber for body armor and vehicle protection, which bids up the price for aerospace and sporting goods manufacturers who rely on the same material for everything from airplane wings to tennis rackets.[1] Little remarked upon is the connective tissue that unites these subplots into a bigger story, the dynamic epic of how this element flows through life and industry, entwining evolution with the Earth's inanimate forces, air, sea, rock—and human infrastructure.

We are warned by many books and articles written every year that the carbon cycle, this global flow of carbon, is broken. These crises, sometimes involving life-or-death issues, require continuous public monitoring. Less frequently remarked upon is how the cycle is supposed to work. *The Carbon Age* teases out the wonder of our world of carbon—and the

1

roles evolution and now industrial civilization have played—through times of havoc and times of stability, in the continuously shifting landscape of carbon. Simply stated, the fastest way to learn the most about the world is through the carbon atom.

A reasonable expectation might hold that the ninety-two natural elements mix and match equally to create the infinity of life and the planet life grew up on. Not so. The Universe is reasonable—but not that reasonable. Carbon is the fourth most abundant element in the Universe, but not on Earth. This planet is largely a sphere of oxygen and silicon. Carbon is not even counted among the top ten most abundant elements on Earth.[2] Yet it structures and fuels all of life. Carbon is the citizen king of the elements, performing roles in nature from the menial to the extraordinary, and in so doing governs who we are and what life is. "Carbon's kingliness as an element stems from its mediocrity: it does most things, and it does nothing to extremes, yet by virtue of that moderation it dominates nature," writes Peter Atkins.[3] Why that's the case, the strange case, is the subject of this book.

Carbon is the ubiquitous architect, builder, and most basic building material of life. It's the molecular scaffold of every living creature, and all of the dead ones, too. Carbon is life's quartermaster, delivering stored energy to cells. Every living thing stores its genetic information in the same language, a chemical alphabet written in a carbon script. By mass, carbon is the largest component in the exquisite spiraling staircase of DNA. Some two dozen elements make up all living things, yet 96 percent of the body is comprised of only four elements—carbon, oxygen, hydrogen, and nitrogen—and most of the oxygen and hydrogen is water.[4] Carbon is the Velcro that holds, frees, and remakes the molecules of life. The oil in our hair and the fat in our gut, skin, and other tissues are essentially hydrocarbons, or carbon frames plugged with hydrogen. Sugars contain carbon, hydrogen, and oxygen, and include everything from sucrose, which gives a cookie its sweetness, to deoxyribose, which girds the parallel helices of DNA. Amino acids are built around single carbon atoms and contain all four of life's most abundant elements, at least. They are the building blocks of proteins, "nature's robots,"[5] which structure cells and perform their work.

Carbon is more than life's master builder. Civilization is built on a carbon foundation, one much wider than just the global combustion of fossil fuels. Living things and inanimate objects are more similar than they appear. For example, plastic is not just what soda bottles are made of. Plastics are polymers, chains of repeating units, whether they are found on tortoise shells or tortoiseshell sunglass frames, in chickens or rubber chickens. Chemists point out that nylon is a polyamide, like proteins, but it is composed of a single, repeated synthetic amino acid. "We are, all of us, warehouses of natural plastics," a specialist in plastics failure once wrote, driven by his wife's cancer to see the tragic similarities between chemical degradation and illness.[6]

If carbon is so instrumental to life—and has always been—it's tempting to say that every age, from the foggy beginning, has been one of carbon. Yet only one gets to wear the name tag. The Carboniferous period, which spans from 359 million to 299 million years ago, earned its name from massive burial of coal-forming organic carbon.[7] During that period there was a great accumulation of atmospheric oxygen. Woody trees with leaves thrived during the Carboniferous. Their roots altered earth and accelerated the erosion of rocks, sweeping carbon and minerals to the sea, where marine life absorbed some of them. Some of the carbon eventually settled to the seafloor. As erosion drew terrestrial carbon away, the plants and trees continued to exchange molecular oxygen (O_2) taken from water in return for airborne carbon dioxide (CO_2). Tall, scaly trees grew old and fell into bogs, hiding their carbon underground, where it stayed until the coal industry began unearthing it some three hundred years ago—seconds in geological time.

Carbon structures life. Oxygen ignites it. Evolution in the Carboniferous entered a period of experimentation that wrought 175-foot, scaly trees with long, floppy leaves, and dragonflies with 30-inch wingspans, a world we might recognize only as science fiction.[8] Today's carbon age is the Carboniferous period in reverse—coal-fired power plants burn long-buried carbon back into the atmosphere, recombining carbon and oxygen into CO_2. Your car is doing the same thing. Gasoline is refined from petroleum that leached from source rocks much more "recently" than the Carboniferous. Most oil became trapped in its underground vaults within the last 90 million years. Not surprisingly, the rise in carbon dioxide associated with global warming is accompanied by a (nonthreatening) decline in the amount of atmospheric oxygen.[9]

We have recently become carbon conscious, trying to measure our "carbon footprint" as if it were a shoe size. That's fitting, since we are now trying to manage carbon in a way it has never been managed before—a direct consequence of mismanagement. Our carbon age is marked most visibly by rational human attempts to wrestle the Earth's geochemical cycle back from the brink, after two centuries of burning hydrocarbon minerals in industry.

All of the research, all of the rhetoric, all of the rancor over global warming boils down to two basic notions, ideas that shape parts I and II of this book.

First, the Earth's temperature and the carbon content of the atmosphere are correlated on every geological time scale. When temperature goes up, carbon goes up. When temperature goes down, carbon goes down. That's usually the sequence of events—first temperature, then carbon—until industrialization, when scientists know that the reverse is true.

The geological record suggests that life has always helped regulate the carbon content of the atmosphere, even though geophysical forces predominate over large timescales. Humans can lay claim to being the speediest, and if trends continue, the most self-destructive regulators of climate. Carbon flows through the atmosphere—by mass a tiny fraction of the global cycle—which is frequently thought of as the flow of carbon through living things, the atmosphere, oceans, land, and from hydrocarbon minerals.[10] These are not static categories, "stove pipes" cut off from each other. They are natural conveyers that converge, forge new directions, and vary their speed. The atmosphere is a (geologically) short-term way station for carbon atoms passing out of living tissue, the high seas, or terra firma.

Dynamism marks the Earth's climate cycle, but stasis is everywhere locked into the way we think about, organize, and apply scientific knowledge. The administrative and intellectual categories that we divide our institutions into—congressional committees, high school science curricula, newspaper sections—are dozens or in some cases hundreds of years old, and no longer adequately match how we understand the world to work. Universities are rapidly shifting and redefining disciplinary boundaries, particularly as health and biology adopt tools spun from advances in physics and chemistry. "A number of important questions are only able to be addressed at the nexus of the life sciences with the physical, computational and information sciences," says Shirley Ann Jackson, president of Rensselaer Polytechnic Institute.[11]

Instead of thinking about the world in old-fashioned, static subject categories, such as "climate," "geology," "marine chemistry," or "biology," nonscientists might think of the behavior of carbon from the molecular to global scales, as "carbon science." Instead of keeping track of confusing boundary distinctions among scientific disciplines, just scrub them, and *follow the carbon.*

Science is a curiosity-led, puzzle-solving enterprise. Physical evidence and critical reasoning bring advances in knowledge and the technologies that underpin our lifestyle. Great and aspiring scientists attack the world through the "scientific method," a tool for logical thought applicable far beyond the laboratory, even if it's rarely acknowledged beyond the schoolroom. Scientists organize their findings in a hierarchy of knowledge. Confirmed observations are "facts"—trivial but essential bricks in the larger edifice. Accumulated facts can point to broad patterns within nature. As they begin to make sense together, they might suggest a "hypothesis," which generates testable predictions about the phenomenon under study. A hypothesis that passes many experimental tests can become a scientific "law," which reliably predicts how a phenomenon behaves. A law may be elevated to the highest echelon of scientific knowledge— "theory"—when its explanatory power reveals fundamental workings of nature.[12]

The second observation that launches this work is that humans have sped up the global carbon cycle at least one hundred times faster than usual, transforming the world into one that we eventually might not recognize as our own.[13] Abrupt climate change has occurred in the past, as recently as 12,000 years ago, but scientists see no analogue in the record to suggest what searing changes the future holds, or how "abrupt" change can be.[14] Man-made global warming is a geological aberration, nearly meteoric in speed.

Homo sapiens is not the first species to hobble a period of stability. But a look at how the Earth works leads to the conclusion that we are changing it faster than ever before, as never before. Scientists distinguish between short-term and long-term carbon cycles. The short-term cycle lasts hours, years, or millennia, and describes carbon's path through life, waters, soil, and air. The long-term carbon cycle adds another leg to the path, the

Earth's crust.[15] When material settles to the ground or ocean floor, and stays there, carbon's travel can halt for millions or tens of millions of years. Carbon can remain underground for 200 million or 300 million years, or longer, until a volcano or undersea vent blows it back to the surface system, mostly as carbon dioxide. As a geological phenomenon, human civilization short-circuits the long-term and short-term carbon cycles.

The Earth, its waters, rock, and atmosphere are a massive carbon-laundering operation. Just as carbon atoms flow through regular, chemical cycles within living cells, so, too, do they flow through the grand geochemical cycles of nature. Geologists estimate, and re-estimate, the Earth's carbon reservoir at more than 75 million billion t (metric tons), mostly buried in limestone, dolomite, ossified gunk called kerogen, coal, oil, and natural gas. That's the estimated amount Earth has and has had since its inception. Matter is conserved on Earth, neither created nor destroyed. It just cycles. The atmosphere carries about 900 Gt (gigatons), or billion tons of carbon. That's more carbon than resides in terrestrial plants, which store about 600 Gt. Soils absorb three times that amount. For all its disproportionate influence on Earth's habitability, atmospheric carbon makes perhaps a thousandth of a percent of the Earth's total store. Recoverable fossil fuels might hold between 5,500 and 11,000 Gt of carbon. Most of the short-term carbon flows with the oceans, which carry about 42,000 Gt—maybe fifty times more than the atmosphere.[16] Most of that lies in intermediate and deep waters. The oceans absorb about half of human fossil fuel emissions, a trend that will not continue forever, and that is already showing signs of slowing down.

Wind and water, bacteria, botanicals, and beasts wind through one another, eternally, in the carbon cycle. Carbon atoms hang in the atmosphere for a century or longer, flanked by oxygen atoms, as carbon dioxide (CO_2). CO_2 molecules ricochet off other air molecules, jostle, vibrate, and flip through the atmosphere. Plants suck in this atomic troika from the air and store it temporarily in leaves. An Apatosaurus comes by and eats the leaves. A Tyrannosaurus leaps out and eats the Apatosaurus. The Tyrannosaurus lives to about twenty-eight, before he expels his last lungful of carbon dioxide.[17] Scavengers and bacteria break his corpse down into more carbon dioxide and nitrogen and the two dozen other elements he unwittingly carried around. Rain washes some of his decayed material to the sea, where his carbon can spend 100,000 years in circulation. Everything has its place in the carbon cycle, until it has another place.

It's not only Eastern mystics and Western self-help gurus who preach that everything is in some advanced state of becoming something else. Chemically and biologically, it's true. Life and the atmosphere, oceans, and earth are locked in a billions-year-old dance that will continue until the Earth's volcanic interior shuts down or the Sun burns out, whichever comes first. Until that happens the flow of energy through the Earth system will drive atoms and molecules into ever-increasing complexity.[18]

In our carbon age, economic activity has collapsed geological time into a human life span. The first half of this book explores the origins of carbon and life, and cases in which evolutionary innovations redirected how carbon cycles through the atmosphere, oceans, and land. The second half covers just the last 150 years, and explains how scientists, industrialists, and consumers created what amounts to an industrial carbon cycle—the flushing of millions of years of geological sediment back into the atmosphere. The six chapters of part II mirror chapters in part I, as a way to compare how evolution and human technology address similar problems, technology's place within evolution, and a way to demonstrate something fundamental about how this world of carbon works.

PART I | The Natural

At this stage you must admit that whatever is seen to be sentient is nevertheless composed of atoms that are insentient. The phenomena open to our observation do not contradict this conclusion or conflict with it. Rather they lead us by the hand and compel us to believe that the animate is born, as I maintain, of the insentient.

—Lucretius, Roman poet and philosopher, ca. 100–50 BC

1 | Out of the Frying Pan
Carbon After the Big Bang

Romantics might like to think of themselves as being composed of stardust. Cynics might prefer to think of themselves as nuclear waste.

—Simon Singh

Half god, half man, Prometheus knew his mortal relatives would benefit from the fire deities controlled. He stole it, enraged Zeus, and suffered a unique cruelty. Prometheus stands chained to a mountaintop. Vultures feast on his liver daily. At night his wounds heal, and so it goes, into eternity.

The Prometheus myth has reverberated throughout time, and is particularly apt for our age. The gift of fire is the gift of enlightenment. And a flame, as you would imagine it with your eyes closed, is incandescent carbon. How carbon came to Earth, how humans controlled its burning, and then lost control, is a story Olympic in scale, written and rewritten by thousands of scientists, for hundreds of years, and under revision right now.

Carbon's earliest episodes occur long ago and far away. They are prehistory to the breath and Promethean fires of Earth. Were it not for two favorable conditions, carbon would never have amounted to much, and none of us would be here. What scientists might call "favorable conditions," from a human-centered perspective we might think of as felicitous quirks of the Universe. They are: how stars forge carbon from lighter elements in their cores, and how carbon gains freedom in interstellar space, where its kingliness among the elements first emerges. The story begins like this, millions of years before the first carbon atoms graced the Universe.

There was nothing, and an instant later the Universe screamed into being, opaque with energy raging at the hottest temperatures ever known to occur, hotter than the Sun's 15-million-degree C (27-million-degree F) interior by a factor too large to apprehend. The Universe's birth is marked in time increments too small to experience, infinitesimal fractions of a second that cosmologists call "epochs." Before a millisecond passed, the four physical forces that govern how particles of matter interact extricated themselves one from another: gravity first, then the force that holds atomic nuclei together, the force that governs radiation, and finally electromagnetism, which enables the play of nuclei and their electrons—and by extension the unlimited universe of human experience, of all living things, in our tiny, pulsing world of carbon.[1]

The infant universe grew, but the energy it contains remained the same. Energy cannot be created or destroyed. The Universe has as much energy in it today as it did in its first yoctoseconds; it's just much bigger. As the Universe continued to expand through those early epochs, its energy dissipated through a larger space. Consequently, the temperature dropped. Energy blinked into matter. Positively charged protons, neutral neutrons, and negatively charged electrons are atoms' three main constituents. The lightest atomic nucleus is hydrogen, which in most instances has just one proton, the only element to appear in the minute-old Universe.

Before the first stars formed, long before carbon nuclei appeared, the violence that forges them first appeared in the Universe. A minute or so after the big bang, subatomic particles first welded themselves together into larger nuclei. Protons and neutrons collided and fused, emitting a fraction of their mass as energy. Their union left behind nuclei containing two protons, characteristic of the element helium. Another proton turned helium into lithium. After two minutes or so, the temperature cooled below the threshold required to sustain nuclear fusion. That left hydrogen, some helium, and much fewer lithium nuclei as the only elements. There were no atoms for many thousands of years after the big bang, just "undifferentiated ooze," as astrochemist Eric Herbst describes it, nuclei and free electrons ripping about space.[2] Atoms became distinct about 380,000 years after the big bang, when the temperature permitted positively charged nuclei to capture electrons.[3] The debut of these electrically neutral atoms lifted the ionic haze that enshrouded the Universe. Light streamed across space for the first time.[4]

Over the next 200 million to 400 million years, the atoms concentrated into clouds, drawn by their mutual gravity. The denser the cloud, the stronger its gravity pulled atoms together, and the more atoms joined the mix. Atoms collided with each other with greater frequency and intensity. The collisions excited the electrons, and the temperature rose. Nuclei let go of their electrons as the cloud heated and condensed, heated and condensed, until the core reached a critical temperature, several million degrees kelvin, hot enough to reignite nuclear fusion, the element-building process dormant since the first minutes. Light poured from stars many times larger than our Sun.[5]

The first stars lived fast and died loudly in supernovae, burning for just a few million years, a cosmic flicker. Billions of years later, supernovae are only rare explosions, the spectacular stellar deaths that occur when a stellar core collapses to a fraction of its original size in half a second. The detonation instantaneously releases three hundred times more energy than our Sun will put out in its entire 10-billion-year life span.[6]

The first generation of stars was an anomaly. Most mature stars—including the Sun—live orders of magnitudes longer, and die less dramatically. In the abruptness of their lives and power of their deaths, it's easy to overlook the importance of the ash the first stars launched into space. The first elements in the Universe that were heavier than lithium were carbon, nitrogen, oxygen, and traces of heavier elements.

The big bang dispersed only the three smallest elements, hydrogen, helium, and lithium, which scientists have assigned the atomic numbers one, two, and three. These numbers indicate how many protons are in a nucleus and constitute the signature that defines atoms as one or another element.

The key to how stars build carbon comes from the *mass* of atomic nuclei—the sum of protons and neutrons, called isotopes. Each element has different isotopes, depending on the number of neutrons cohabiting the nucleus with the protons. Hydrogen has one proton, and can have between zero and two neutrons; it has potential isotopes of one, two, or three. Helium has two protons and can have one or two neutrons, which contribute to its isotopes of three or four. Carbon is built by the isotopes of hydrogen, indirectly, and helium, directly.

The heat and pressure of stellar cores bake hydrogen into helium, changing an entire star's composition one reaction at a time. Two hydrogen-1 nuclei fuse, and their masses add up to a larger isotope. A third hydrogen-1 fuses with hydrogen-2, which is called deuterium. Every new fusion reaction unleashes energy, enough to compensate for the mass difference between the two original nuclei and their product. Energy and mass can be expressed in terms of one another. This fact is embedded in Albert Einstein's famous equation, $E = mc^2$—energy equals mass times the speed of light squared (or more than 186,000 miles per second, squared).

Nuclear addition skips a few beats after these reactions. Those hydrogen-3 (tritium) nuclei can't become hydrogen-4, but they can do this: run at each other and fall apart. When two hydrogen-3 nuclei react, they fracture instantaneously into a brand-new helium-4 nucleus and two leftover hydrogen-1 nuclei.[7] That's what happens inside stellar cores and how young stars burn, by making helium-4, the direct nuclear ingredient for carbon.

Stars are so large and dense that energy released in their cores requires some 200,000 years to reach the surface. The energy is continually absorbed and emitted by other particles along the way, up through convection layers, to the surface, past the flares arching a million miles high, and out into space. Stars need billions of years before all this burning—hydrogen to deuterium to helium—transforms their composition. Helium's mass is greater than hydrogen's. Its heavier nuclei exert stronger gravity. Over the Sun's history, helium's gravitational pull has caused the orb to contract, generating more heat. The Sun was about 30 percent cooler when life began on Earth, about 4 billion years ago, an observation that sets a parameter for thinking about the earliest conditions that life emerged under.[8]

The elements just larger than helium—lithium, beryllium, and boron—are too rare in the Universe to play a significant role in stars' energy production. So when conditions are just right, the star large enough, and its core hot enough, helium begins to burn into the next largest element: carbon.

Orderly scientific narratives, such as this tale of the prehistory of carbon, grow by accretion, over many decades, as hypotheses compete for attention and scientists hunt for evidence to support them. Scientists are people. They have biases, career pressures, and make errors. But science, as a global, centuries-old enterprise, is supposed to eliminate individual biases and correct errors, creating a self-doubting, often contentious, accretive body of work based on physical evidence and critical reasoning. Over time, the professional community separates promising ideas from ones unsupported by evidence.

Sometimes professionals become waylaid. In 1900, some physicists felt they were running out of problems to solve, even though no one knew how the Sun shines. They knew how it didn't shine. If the Sun burned like a candle, its fuel would have vanished shortly after its birth—very shortly. No practical amount of combustion can match stellar nuclear fusion. A million billion tons of coal shoveled into a furnace would keep pace with the Sun's power for little more than a second.[9] That much coal, if it existed, could theoretically fuel an economy equal in size to that of the United States for 1.5 million years.

The discovery of how stars make carbon played out as a side drama within the search for the origin of the Universe. Stars produce larger elements as they age, which intrinsically ties pursuit of their energy production to element building. And where the elements came from was central to considering the origin of the Universe. Carbon's role in nuclear fusion turned out to be a major battleground when the big bang theory was still the big bang hypothesis.

Also in 1900, German physicist Max Planck proposed a transformative idea that caused the entire field of physics to pivot: energy moves discontinuously. Subatomic particles—even particles of light—do not fly, beam, or buzz. Particles are not like sponges, soaking up energy like water and then squeezing it out. Sometimes energy levels are depicted as stairs; you have to be on one or the other.[10] Particles might skip, if skipping means changing their location from one point to another without traversing space. These increments are called "quanta." Quantum is a Latin word related to the English word "quantity."

Quantum physics grew up during the next four decades, which were filled with enormous creativity, productive failure, theoretical insight, infighting, dense mathematics, unproductive failure, and most important, experimental evidence. Cornell scientist Hans Bethe helped canonize the formative period of nuclear physics with a series of encyclopedic articles in the mid-1930s that became known as "Bethe's Bible." That primed him for further achievement. In 1938, nuclear physicists' peers began to ask serious questions about the Sun's power. Early that year, Bethe published a paper with a younger colleague about the most basic nuclear fusion reaction, of two protons, or hydrogen nuclei.[11] In so doing they drew on the physics breakthroughs of previous decades and inaugurated the theoretical framework for research programs that would last the next several decades.

After an annual physicists' conference in Washington that year, Bethe realized that there is a second way that stars burn hydrogen into helium—but they need carbon to do it. He described what he called the "carbon-nitrogen-oxygen (CNO) cycle," a phase that stars larger than the Sun go through. In this cycle, a carbon nucleus and four hydrogen nuclei (or protons) fuse in a series of reactions that produce a new helium-4 nucleus and the original carbon-12 nucleus. Carbon returns to start the cycle fresh with four more protons.

Despite the significance of this work, a colleague couldn't resist the temptation to make it appear even more epic. A ringleader in the nuclear physics community, George Gamow of George Washington University, fibbed that Bethe's insight was all the more heroic because he had it on the train home to Ithaca, New York, after the scientific conference in Washington (that Gamow cohosted). Bethe's colleagues held him in such esteem that the story seemed plausible, and stuck for a long time. (In fact, it took him a few weeks.)[12] Bethe's study was a pivotal moment in the sixty years that passed from Max Planck's discovery to the first thorough study of how stars build all of the elements, in 1957.[13]

This view of the stellar carbon cycle holds today, even after seventy years of fine-tuning. Based on then-current solar physics, Bethe calculated that this carbon burning occurs in the Sun. This gave Bethe the impression that the Sun had a carbon cycle of its own. In a glib moment speaking to the New York Times about the cyclical nature of the CNO cycle, Bethe told the paper, "The Sun eats its carbon and has it, too."[14] Ambiguity over which cycle the Sun uses—the "proton-proton" chain or the carbon cycle—persisted into the early 1950s.[15]

Neither Bethe nor anyone else foresaw that carbon would become a pivotal, but not decisive question in the larger fight over the Universe's origin. Bethe's work left open the question, "Whence carbon?" He tried to answer it. He was unable to, perhaps because key nuclear physics experiments would not be performed for another decade. He dismissed the idea that three helium-4 nuclei fuse instantaneously into one carbon-12.

All the players who launched the big bang hypothesis and the synthesis of the elements in stars were monumentally right and wrong about various things. The golden age of cosmology occurred in the decade after the end of World War II. George Gamow was a hard-drinking Soviet émigré to the United States with a brilliant mind, an ebullient sense of humor, and a vision for bringing maturity to a then-young field. In 1948, Gamow and graduate student Ralph Alpher sketched out the basics of their theory of origins. They posited that the energy released during the Universe's initial expansion grew cooler and cooler, from many billions of degrees in the earliest epochs to about −268°C (about −455°F) today, 14 billion years later. This comes out to 5°K (kelvin), the absolute temperature scale preferred by astronomers.[16] That estimate, 5°K, was an astonishingly bold prediction that within twenty years turned out to match astronomical observations with quickly forgivable imprecision.[17] But at the time of its articulation, Gamow's hypothesis earned scorn from some prominent astronomers and nuclear physicists. "Many of us, theorists especially, were antipope enough to not like the creation of the universe in a sudden explosion," said Edwin Salpeter, a Cornell astrophysicist and Nazi-era Austrian refugee. Gamow's formidable, iconoclastic opponent, Fred Hoyle, ridiculed Gamow's idea as a "big bang" at the origin of everything, during a BBC radio show about science. Gamow loved Hoyle's derisive phrase and the name stuck.[18]

Gamow and Alpher suggested that during the big bang, fusion built smaller nuclei into larger ones a single proton at a time. The trouble is that nuclear fusion hits roadblocks. Experimental observations already undermined this idea. In 1939, physicists at the University of Minnesota demonstrated that lithium-5 is unstable and breaks up quickly. Willy Fowler's lab at the California Institute of Technology (Caltech) found a decade later that mass 8, a beryllium isotope, had no stable nucleus either.[19] Carbon's main isotope has a mass of 12. But Gamow and Alpher's hypothesis couldn't account for nuclei heavier than helium-4.

In the early 1950s, nuclear physicists at Caltech looked deeper into the instability at beryllium-8, and found a curiosity, one that shed light on the creation of carbon in stars. The university's Kellogg Radiation Laboratory housed some of the top minds in the business. Under the direction of

Fowler, Caltech scientists conducted some of the most comprehensive experiments in nuclear physics, their electrostatic generators beaming particles into each other to reveal their makeup. Edwin Salpeter, then a young Cornell professor visiting Caltech, picked up on Hans Bethe's ruminations about how some stars might release energy by burning three helium-4 nuclei into carbon-12. Physicists call these helium-4 nuclei "alpha particles." (They also make up the radiation that uranium emits over its long half-life and that americium emits in a functioning smoke detector.)

Salpeter saw something new in the 1949 Caltech studies on beryllium-8, the union of two alpha particles. He recognized this fleeting nucleus as a pivotal stepping stone to carbon, one of the "quirks" that explains carbon's abundance and therefore enables the possibility of life as we know it. This beryllium-8 nucleus exists in its ground state for 0.968×10^{-16} seconds before breaking up. Only astrophysicists could describe 0.0000000000000000968 seconds as "sufficiently long lived"[20]—but that's how long beryllium-8 holds, and it is long enough to unite with a third alpha particle to create carbon. If it doesn't, the beryllium splits in half. Despite the nucleus' very short life span, stars make it in ample quantities— about one nucleus in 10 billion particles—that carbon-12 has a favorable chance.[21] Consequently, the burning of beryllium-8 and helium-4 into carbon-12 became known as the "triple-alpha process," or more familiarly, the "Salpeter process."

That's not the whole story, of course. This research answered a pressing question even as it posed a substantial problem. If true, stars would produce carbon too slowly, and in insufficient quantity to match astronomical observations. Enter Fred Hoyle, coiner of "big bang" and proponent of a then-competitive hypothesis for the Universe's origin. Hoyle could make sense of the amount of carbon observed in the Universe only if a previously undiscovered property lurked within the carbon nucleus, an energy level close to the addition of the energy levels of beryllium-8 and helium-4. Once reached, this level must allow three alpha particles to lock into a carbon-12 nucleus and stop there.

On a yearlong visit to Caltech, Hoyle pressed the Kellogg Radiation Lab into testing whether or not his predicted energy level in the carbon nucleus actually existed. "[Willy] Fowler has said, in later years, that his first impression was that I had somehow gone a long way off my mental compass bearings," Hoyle wrote.[22] But experiments proved his prediction to be correct. When helium-4 and beryllium-8 nuclei fuse, their combined energy is incrementally *less* than Hoyle's predicted level in carbon, at 7.65 million electron volts. The star's heat pushes the beryllium-helium combination to this predicted level in carbon-12 and the reaction sticks. If a fourth helium nucleus approaches the new carbon nucleus, potentially to make oxygen, the energy of that reaction is incrementally *more* than the nearest oxygen level, even before accounting for the star's heat. The reaction fails. Carbon and helium retain their separate identities.[23]

Caltech nuclear physicists found Hoyle's predicted level in carbon in 1953 after ten days of experiments. Hoyle likened that week and a half to an accused criminal's anticipation of the jury's return—with a key difference. The defendant knows if he is guilty or not, and hopes for exoneration in any case. "In physics, on the other hand, the jury of experimentalists can be taken always to be right. The problem is that you don't know whether you're innocent or guilty." It took a few years to convince distinguished holdouts.[24]

Hoyle's prediction and the experiments that confirmed it gave credibility to the triple-alpha model of carbon's origin. Even better would be to fuse three alpha particles into carbon in a laboratory experiment. However, scientists have no place to truly test the carbon-making process; the core of the Sun is too cold a place to conduct an experiment. Hydrogen bombs fuse protons together and reach "only" several million degrees. Yet the evidence for how stars burn helium into carbon advances nonetheless. The triple-alpha investigations that began in the early 1950s were supplemented recently by a team of European physicists using two particle accelerators, in Switzerland and Finland. They didn't make carbon from three helium-4 nuclei the way some stars do. But they did manage the reverse; breaking down carbon-12 nuclei into three alpha particles,[25] and in so doing peered into the 100 million°K core of large stars, the birthplace of the carbon within and around us.

In his later years, as more than a few peers would claim,[26] Hoyle did go "a long way off his mental compass bearings." He professed a rabid true-believer's atheism. Among his more mundane idiosyncrasies, he ascribed intent to the particles he studied, an unusual and dubious position for a professional scientist. He wrote that the only intelligent—rather "superintelligent"—explanation of the carbon atom's properties is that it was engineered by forces striving for perfection. "Otherwise, the chance of my finding such an atom through the blind forces of nature would be less than 1 part in $10^{40,000}$," he wrote. "If you were a sensible superintellect you would conclude that the carbon atom is a fix."[27]

In 1953, Hoyle's motivation was to prove George Gamow's big bang hypothesis incorrect by showing that nuclear fusion by the addition of a proton at a time wouldn't work. He argued instead that the Universe has always existed, in a "steady state," and that stars themselves make the elements, not some big bang at the beginning of time. His and Salpeter's work on the origin of carbon essentially cut off Gamow's original hypothesis at the knees, showing that stars, not the Universe's birth, produced elements heavier than hydrogen, helium, and lithium. The triple-alpha process is the likely route by which carbon came to be. But from here, the story of carbon can branch out into an ever-increasing infinity of directions. "The number of atoms is so great that one could always be found whose story coincides with any capriciously invented story," wrote Primo Levi in an essay about carbon.[28] This story now turns to a likely way by which carbon's properties became liberated, a favorable condition—or quirk of the Universe if you prefer—that raises carbon's achievements up a notch.

The late Caltech physicist Richard Feynman mused that the single most important scientific statement to save, in the unlikely event that all other knowledge be destroyed is, "All things are made of atoms—little particles that move around in perpetual motion, attracting each other when they are a little distance apart, but repelling upon being squeezed into one another."[29] Nuclei regain electrons and become atoms in the cooling outer envelope of dying stars. When their attraction and repulsion balance, atoms collapse into molecules.

At a star's death, hydrogen and helium blow off into space. Extraordinary carbon molecules and amorphous soot form outside a star's cooling outer envelope: diamonds nanometers in size; molecular carbon rings, horns, and balls; carbon monoxide in great quantity; graphite; long carbon chains flanked by hydrogen on one end, nitrogen on the other; acetylene (an industrial feedstock and a gas used in welding closer to home); and many other familiar and foreign basic chemicals.[30] Gases glow ethereally and take on fantastic shapes, gently distributing their remains to the Milky Way Galaxy.[31] Freed from the star, gas and dust can travel 100 million years before settling down again.

The astronomical cycle begins anew as gravity pulls matter together into interstellar clouds—gargantuan incubators—where carbon's architectural potential emerges. "Diffuse clouds" contain pockets of "dense interstellar clouds"—a relative term since at 10,000 molecules per cc (cubic centimeter), the gas is far closer to a vacuum than even the emptiest laboratory vacuum.[32] Much of the Galaxy's oxygen, which outnumbers carbon by about 2.3 to 1, may be locked up in dust grains scattered throughout the clouds that perform two important tasks. The dust shields the cloud's interior from destructive ultraviolet radiation that would fry the molecules forming on the inside. Second, it helps build the most abundant molecule in the Galaxy, paired hydrogen atoms, H_2.[33] Carbon's architectural potential becomes visible for the first time in these 10°K clouds (−442°F).

Carbon owes its opportunity to become citizen-king of the elements to aloofness between hydrogen molecules and helium atoms in these cold clouds. If helium and hydrogen molecules didn't attack it, carbon might remain locked in carbon monoxide and its creativity never have a chance to flourish. The carbon in our bodies and modern materials, the carbon that sat underground for hundreds of millions of years and has burned into atmospheric CO_2, may once have been locked in carbon monoxide in dense clouds and freed because of this electronic barrier between helium and hydrogen.

About 1,600 light years from Earth, the Horsehead nebula is a dense cloud of gas in the constellation Orion.

Hydrogen, carbon, and the other main atoms of life bond covalently, which means they share electrons. Electrons inhabit shells around the nucleus, and they each occupy their own "orbitals" within these shells. Most atoms do not have the maximum number of electrons possible in their outer shells. In covalent bonding, atoms conjoin, so that each atom fills up those shells with shared electrons. Every molecular bond is a balance between two forces, the attraction between protons and electrons, and the confinement energy required to keep each electron inside its own territory.

Carbon has six electrons to complement its six protons. Six neutrons, plus those protons, give its main isotope a mass of 12 (The nucleic particles are more than 1,800 times more massive than electrons.) But the way atoms work, only the outermost electrons participate directly in molecule making. Carbon has six electrons, but only four can bond. They do so prodigiously, but not irreversibly. Carbon makes good strong bonds, up to four of them at a time. But it breaks them and forms new ones, too. This combination of strength and fragility enables all of carbon science and all of life.

Carbon's success comes at what seems like unbearable odds against its relevance. First, the Galaxy is mostly empty space. Second, where there are clumps of matter—clouds and stars mostly—nearly all of it is hydrogen and helium. And third, that's only the 4.4 percent of the Galaxy—indeed the Universe—that we understand. Dark energy, the force that makes galaxies accelerate their expansion from each other, makes up 73 percent of the Universe. Dark matter, which accounts for gravitational forces on stars and galaxies, constitutes 23 percent. It is called dark because it does not interact with any light wavelength on the electromagnetic spectrum, from low-energy radio waves to high-energy gamma rays. As Sean M. Carroll of Caltech quips of dark matter, "Most of the Universe can't even be bothered to interact with you."[34]

So, carbon makes up just three ten-thousandths of 4.4 percent of the Universe. That doesn't sound like a very promising starting place. Yet it must be. Carbon appears in 90 percent of molecule types detected between stars. Since the late 1960s, astronomers have spotted energy signatures of more than 130 kinds of molecules in the dense clouds, such as the Orion nebula, about 1,600 light-years away from the Earth, and Sagittarius B2, near the Galaxy's center. Experimentalists reproduce these signals in simulated conditions in laboratories to make sure radio telescope readings are matched to the right molecules. Some of these molecules are familiar enough on Earth. For example, there may be enough ethyl alcohol in a dense cloud to fill one hundred shots of vodka each the size of Earth.[35]

Astronomers understand what carbon is doing up there because gas molecules take in and radiate detectable energy at signature frequencies. Electrons become excited when molecules collide or the right wavelength of energy zips in. When they get excited they "leap" to a higher energy level. After a time molecules re-emit it. For example, carbon-nitrogen molecules called "cyanogens" are sensitive to 2.73 K energy, about the same frequency as the big bang's universal echo. Astronomers detected cyanogens in five "diffuse clouds" in the early 1990s. These observations added proxy evidence to direct physical measurements made by two Bell Labs scientists in the 1960s and by NASA in 1992.[36]

These telescopes have detected signatures from dozens of molecules since the 1950s, including the cyanogens in the early 1990s. Gas molecules pick up radio-frequency energy, which is much less powerful than white or visible light, the target of more familiar optical telescopes and the eyes that peer through them. The light that we see is a fraction, 4 percent, of the electromagnetic spectrum. This span includes gamma rays, x-rays, and ultraviolet light on the high-energy, high-frequency end, and infrared, microwave, and radio wavelengths on the other. Inside a molecule, electrons "skip" up or down energy levels when they absorb or emit radiation. While this interaction of energy and molecules may seem an arcane detail of radio astronomy, it is also fundamental to understanding everything from how plants convert sunlight into carbohydrates, to how animals see, to how greenhouse gases absorb heat.

Correlations among telescope observations, experiments that reveal specific molecules' molecular signatures, and understanding of the general behavior of atoms give scientists confidence in this story of how enmity between hydrogen molecules and helium atoms free carbon's architectural creativity. It takes more than 24 V (volts) of electricity to jolt helium into dropping an electron. Hydrogen holds on less tightly, letting go after 14 V—a little more than your average 12 V car battery. Carbon, the "friendliest" of the elements, lets go one of its six electrons at 11 V.[37]

Cloud dust blocks ultraviolet light, preventing it from destroying the chemical incubator within. But it cannot keep out cosmic rays—protons ripping uniformly around the Galaxy. When cosmic rays tear through clouds, they free electrons from their confinement. Even helium fails to protect its electrons from cosmic rays. They zoom through with 100 million V, instantaneously ripping electrons from molecules and atoms. Molecular hydrogen (H_2) and helium, which make up nearly all of the clouds, are proportionally hit. Hydrogen and helium become positively charged particles, which are called ions, written as H_2^+ and He^+. Their mutual neglect unlocks carbon's potential.

Molecular hydrogen is mostly off-limits to helium ions seeking electrons. Helium ions react with H_2 ten thousand times less frequently than scientists expect ions and molecules should.[38] Instead, the ions attack the next most available electron source, carbon monoxide.

Carbon monoxide may indirectly be the Ur-molecule of life, the original feedstock for carbon molecules larger than it. There is so much CO in the dense clouds, radio telescopes map the skies based on where it is.[39] Living things may owe their existence to its dissolution. On Earth, you would never know how important this molecule is by the way people treat it. That's because carbon monoxide kills us rather quickly. In the dense cloud it is the prey.

As strange as it sounds, molecules take "baby steps" toward familiar substances in the dense clouds. An ionized hydrogen molecule (H_2^+) cannibalizes other H_2 molecules, making a hydrogen atom (H) and a larger hydrogen ion (H_3^+). The latter ion latches on to carbon monoxide, making a hydrogen molecule and HCO^+, an important building block for the complex prebiotic molecules in space.

Helium atoms outnumber carbon monoxide molecules by about five hundred times in the dense clouds. When the cosmic rays break neutral helium atoms into ions, the latter steal carbon monoxide electrons, busting the molecules into neutral oxygen atoms and carbon ions (C^+). That's a big deal. That's how the fourth most abundant element in the Universe, which makes up just a tiny percentage of matter, realizes its potential to build 90 percent of the known molecules in dense clouds—and down the road, more than 99 percent of the variety of substances on Earth.

Carbon locks arms with hydrogen and nitrogen to become hydrogen cyanide (HCN), a deadly poison to humans and, as we'll see, a chemical potentially instrumental to the origin of life on Earth. Free carbon ions barge in on HCN and make HC_3N. Two more make HC_5N, and more Cs hop in until they reach $HC_{11}N$, the longest chain observed in space.

(HC$_{13}$N has been made in labs that simulate interstellar cloud conditions.)[40] These molecules are called cyanopolyynes (cy-a-no-POL-ly-ines) and will "fall to Earth" before carbon's tale is told. C$^+$ builds much more than these straight chains. It builds triangles, hoops, and the Tinkertoy structures of basic carbon science.

The carbon-friendly conditions of dense clouds grow more complicated as pockets of gas and dust condense further into astronomical bodies called hot molecular cores. Molecules continue to fracture and recombine as density increases thousands of times and the temperature rises. The cores pull in rich arrays of molecules, combining them into even more complicated carbon compounds. A long-standing effort of research on hot cores is to find molecules in interstellar space similar to critical building blocks of living things, and by extension, insight into the origins of life.[41]

Amino acids are the building blocks of proteins, and consequently the bulk of organisms' nonwater weight. Life prefers just twenty specific amino acids, although their definition leaves open the possibility for countless more. All amino acids have a carbon atom at their core. Two of carbon's four electrons hold together an oxygen-hydrogen group. A third carbon bond reels in the amino group, made of nitrogen and hydrogen. The carbon's fourth bonding electron ties in a signature side chain, unique to each kind of molecule. All of life builds its hundreds of thousands of proteins with just twenty kinds of these "bricks,"[41] bricks that would never hold together without the aid of a carbon atom.

Astronomers have at least twice claimed to spot glycine, the simplest amino acid, in hot cores.[42] Both were false alarms. These molecules might turn up in cores. They might not. Lack of discovery does not prove they are not out there. In science, ambiguity often reigns. Scientists try not to get too emotionally attached to their results; theirs is a discipline whose main tool is laser-pointed skepticism. If a scientist can't disprove his or her own result, there's more than a lingering possibility that someone else will. "It's not a good idea to take one's science too seriously," Ed Salpeter said, regarding how to think about one's own research. "You need a light touch."

Gravity pulls together the dust and molecules within the hot molecular core, overpowering cosmic forces that work against it. Dust swirls into a gaseous ball. Within, a new star ignites, its rotation flattening matter out into a nascent solar system.

2 | Dancers and the Dance
The Origins of Life

> *Men think epilepsy divine, merely because they do not understand it. But if they called everything divine that they do not understand, why, there would be no end of divine things.*
>
> —Hippocrates

The origin of life has been the subject of myth and religion as long as humanity has told stories around the hearth. But only in the last fifty years have scientists carried out precise experiments to test how the conditions that likely existed on early Earth may have assembled the elegant puzzle of living chemistry. Origins researchers "follow the carbon," from its arrival during Earth's formation to the atoms that gird the genetic code, testing potential pathways through which geochemistry may have sprouted biochemistry.

All life is a unified chemical phenomenon. Yet how life became distinguished from geochemistry is almost a secondary question. Scientists have enough trouble describing what life is in the first place. A human being is as paradoxical as any other organism. Each living, breathing person has untold trillion cells, each cell an elaborate choreography of untold trillion molecules, each molecule a completely lifeless chemical structure, on its own helpless to do anything but wash away or degrade in sunlight. The characteristics we recognize as "life" emerge from these molecules' complex physical interactions, driven by the energy and nutrients they continuously take in from and release to the environment.

Thumbnail descriptions of life usually include the same handful of characteristics. Life is a system of chemicals and self-perpetuating chemical change enveloped in a membrane. These carbon-based chemicals have a love-hate relationship with water. Life takes advantage of energy sources on Earth virtually wherever they are found. And life's diversity emerged because random mutations in genetic coding underwent Darwinian natural selection.[1]

Origin-of-life research demands expertise across many professional disciplines, from geophysicists who study the planet's formation to molecular biologists current in the vast and expanding library of genetics research. Origin research is a worldwide collaborative exercise that falls under the wider rubric of astrobiology, the study of the "origin, evolution, distribution, and future" of life in the Universe.[2] This chapter will survey how scientists from these various disciplines assemble evidence—and frequently fight over the lack thereof—for how living matter evolved from starter chemicals.

Many experimental scientists and theoreticians try to understand how life's ingredients came together into the earliest cells. Their work is sometimes divided into "bottom-up" and "top-down" approaches. The former group considers likely conditions on early Earth, from the atmosphere to deep-sea volcanic vents, and tries to discern how lifeless material may have evolved into living cells. This "bottom-up" approach hits limitations. No lab studying the movement of geochemistry toward biochemistry has been able to synthesize under reasonable conditions anything as complicated as even fundamental constituents of the genetic code, called nucleotides. Each nucleotide is composed of three parts. The "nitrogenous bases" are actually rings of carbon and nitrogen. Carbon rings are so ubiquitous in nature, C atoms are assumed to be there. These are distinguished by having nitrogen in them, too, hence the emphasis on nitrogenous. These bases, adorned with side groups, connect to five-carbon sugars, which with a phosphorous-oxygen molecule form the complete nucleotide. Scientists hit a "black box," a lacuna in knowledge they can't describe despite understanding more or less what fed into it and what evolved out of it.[3]

Scientists inclined toward the "top-down" approach have likened the study of life's universal genetics and metabolism to reading the original text on a palimpsest—an ancient scroll erased and written over several times.[4] This latter group also reaches a black box, from the opposite direction. "You do have your moments when you wonder if this is a solvable problem," Alan Schwartz once said.[5] Problems like this don't have to be immediately solvable in any concrete sense. They only have to be "interesting," a word erased of meaning through overuse in modern standard English, but deployed frequently in scientific papers in a serious meaning—deserving of effort, time, and expense. As a result, the black box shrinks every year.

This top-down approach offers a simple framework for thinking about the penumbra between a sterile world and one inhabited by the first cells. Drawing on previous work, theoretical evolutionary biologist Eörs Szathmáry has laid out three necessities for life: a genetic code or "template" for design information; a metabolism, which converts ingested energy and nutrients into essential biomolecules; and a membrane or "boundary" to concentrate a cell's internal chemistry and regulate what crosses in from the environment. Life must have all three properties, Szathmáry argues, but Darwinian evolution—imperfect reproduction and natural selection—might work on just two of them.[6]

Organisms from the tiniest nanobacterium to the blue whale all ingest carbon to fuel and build these three interlocking systems. Their metabolisms cook carbon-rich "food" into the molecules organisms need to live. Carbon binds with hydrogen, oxygen, nitrogen, and phosphorous into DNA, which is mostly carbon by mass. Carbon girds the membranes or skins that separate their living chemistry from the outside world.

A theme that will run through the rest of part I is that as life became more diverse, and the amount of living matter in the world swelled, the course of its evolution both influenced and was influenced by the global carbon cycle. Since its inception, living matter has helped regulate the amount of carbon in the atmosphere and seas and on land, conditions that in turn influence evolution. The Earth is essentially a closed material system. The amount of carbon, water, and other materials is probably about the same as when the planet formed. From this perspective, evolution is an expanding-contracting regulator of the path of carbon through the Earth system, rewiring the innumerable circuits of the geochemical cycle: the atmosphere, the oceans, and the land. This observation will help set the context for the uniqueness of human development and man-made, or anthropogenic, climate change—the chronology through part II.

Before humans and our fire, before the animals, plants, and microbes, before even genes, metabolism, and membranes, the first prerequisite for life on Earth was Earth. The planet emerged in violence. There's no foundation stone with the date chiseled in roman numerals, but meteorites and the Earth's most ancient lead ores harbor evidence that the planet was born more than 4.5 billion years ago. Both Earth and the meteorites—those meteors that have left traces on the surface of the planet over the years—formed with a certain amount of the radioactive element uranium. Over time, much of it has decayed to form isotopes of lead. Physicists understand how quickly uranium decays, and which uranium isotopes decay into which lead ones. By measuring the modern ratios of lead isotopes, they can count backward to a time when no uranium in the Earth and meteorites had yet decayed into lead—the birth of the planet.[7]

Had nuclear bombs existed back then, they would have been less destructive than the meteors, asteroids, and comets that slammed into Earth, even after it emerged with infant territorial sovereignty. They broadsided the planet for hundreds of millions of years. Meteors the size of mountains rained down every few millennia, making the planet glow like a faint star. The bulk of Earth's carbon may have arrived in carbides—molecules of carbon and a metal. This material would have dissolved in molten iron, its carbon escaping to the surface as gaseous carbon dioxide. The bombardments liberated far more energy than the Sun provided.[8] Our star was young, and accordingly, not so bright.

Sterilized and scalded, sprinkled with volatile chemicals and rocked by forces to shake them up, Earth joined the ranks of planets. Raw materials separated from each other, a violent process that cooked the planet inside out. Iron seeped from the rocky mash and sank to the core. Radioactive elements switched on a natural nuclear-fission reactor that heats the Earth's interior and has kept continental plates sliding about the planet's crust ever since.[9] Oxygen and silicon dominate the Earth's mass, making up about 46 percent and 28 percent respectively, much of it bound together in silicates such as olivine, a mineral of magnesium, iron, silicon, and oxygen.

They were followed by lesser amounts of uranium, lead, calcium, sodium, and trace amounts of others.[10] A hydrogen atom has a tiny mass, even by atomic standards, which obscures its ubiquity and influence. Carbon doesn't make the top-ten list of earthly elements, yet it appears in all but about 100,000 of the more than 33 million known varieties of substances.[11] Carbon is so ubiquitous that chemists don't even mark it with a C in chemical diagrams. The unlabeled vertices in molecular diagrams are carbon atoms.

Geologists call the first half-billion years of the Earth's history the Hadean epoch, for the Greek deity of the underworld and afterlife. This is an odd name in the sense that this Hadean likely predates life, unlike the Hadean domain of myth, which greets souls afterward. Life is inextricable from the solar energy, geothermal heat, movement of the continents, the slosh and spill of oceans, and other hallmarks of the Earth system. And geology had to settle down before life emerged. Plate tectonics is the most significant sustainer of life other than the Sun. It lifts and levels mountains, swallows the ocean floor, opens and seals waterways, and refreshes the feedstock of greenhouse gases and atoms essential to life.

Sunlight and the planet's continuous recycling of surface rock outweigh all other influences on evolutionary change, with the exceptions of the occasional meteorite impact and the geologically unique case of man-made global warming. The Sun, the planet's interior, gradients of acidity or salinity, and radiation all provide Earth with abundant energy sources. Life may have emerged as a kind of chemical steam valve to help Earth efficiently degrade and release this energy.[12]

Bolides, a scientific term for huge objects that hit the Earth, added incremental mass to the growing planet as the more catastrophic phases of planet making waned. Comets shuttled in ice and organic materials from the molecular clouds.[13] Upon impact they vaporized and contributed carbon dioxide and water to the Earth's starter atmosphere. Bombardments punched holes in the young planet, an escape valve for then-subterranean carbon dioxide, which gave the Earth a heat-trapping atmosphere nearly since its inception.

Though the "greenhouse effect" sometimes seems entwined with man-made global warming, it is a universal phenomenon that explains how the Earth has kept sufficiently warm for liquid water almost since its dawn. Carbon dioxide is crucial to the biosphere. As an atmospheric gas, it absorbs heat, blocking its exit back to space. The heat energizes the molecules, making them shake, twist, and flop about. The greenhouse functions on the same principle that allows astronomers to detect molecules in interstellar clouds. The wavelengths CO_2 absorbs are much higher energy than in those dense interstellar clouds, but the principle is the same. Molecules in a gaseous state are like radio antennae. Every station, or molecule, receives a precise frequency. In the atmosphere, CO_2 tunes in to an infrared frequency. Carbon dioxide is a linear molecule, one carbon atom wedged between two oxygen atoms. NASA has named its carbon dioxide monitoring satellite the OCO mission, a triple pun. The official name is the Orbiting Carbon Observatory, but OCO could also practically be a picture of the molecule; and for astronomer polyglots, *oko* means "eye" in Polish.

Earth was fortunate to have a good location and a good greenhouse. Today, the Earth's temperature averages about 14°C (57°F), and is predicted to rise 3.5 or 4°C by the end of this century, enough to transform the planet beyond recognition. Without the greenhouse it might plunge to -19°C (-2°F).[14] For oceans to remain liquid under the faint young Sun, the atmosphere must have provided a greenhouse so powerful that today it would fry some enormous percentage of species into extinction. Earth started out with an exquisite balance, far enough away from the young Sun that it didn't boil, like Venus, yet close enough—and with a strong enough atmosphere—that it didn't freeze, like Mars. Accordingly, we ended up with waterfront property.

Bombardments slowed. Rain poured from the sky and washed into the earliest oceans.[15] Evidence exists that temperatures cooled after the bombardments, making this scenario possible. Microscopic zircon crystals found in Jack Hills, Australia, date to 4.4 billion or 4.3 billion years ago. Geologists determined their age from uranium-to-lead ratios, the Earth's oldest timepieces. The temperature reading came from oxygen isotopes in adjacent rock, which suggests that the rocks formed in contact with liquid water.[16] Rocks this old are rare leftovers. The Earth's first magma oceans or continents long ago sank back beneath the crust, subducted and transformed, probably several times over. Conditions settled for a long time, maybe 400 million years, despite a late resurgence of impacts.

Whatever the atmospheric composition, whether events occurred at

*Charles Darwin, photographed in 1878, nearly two decades
after the first publication of* Origin of Species.

the boundaries between air and sea, sea and rock, rock and air, some combination of these, or deep in the Earth's hot lower crust, somewhere,
somehow, in the unknowable past, carbon atoms first settled with hydrogen, nitrogen, phosphorous, sulfur, and oxygen in "some warm little
pond." Charles Darwin conjured this setting for his friend Thomas Hooker
in an 1871 letter, possibly the most frequently reproduced paragraph in
the history of private correspondence.

> But if (And Oh! What a big if!) we could conceive in some warm
> little pond, with all sorts of ammonia and phosphoritic salts, light,
> heat, electricity &c., present, that a proteine compound was chem-
> ically formed ready to undergo still more complex changes, at the
> present day such matter would be instantly absorbed, which
> would not have been the case before living creatures were found.[17]

Darwin's letter introduces many of the elements still driving much of
modern origin-of-life study: amino acids, phosphates, heat, light, "&c." The

pond imagery stuck, and eventually simmered into the "primeval soup" hypothesis. Darwin was such a good writer, and acquired such personal authority, it's worth casually asking if this metaphor became standard because Darwin wrote it or because he was actually on to something. On a physical level, how else could it have happened?

Time enshrouds life's origins. How carbon atoms zip themselves and other elements into diverse, simple molecules is clear. Some carbon molecules essential to life assemble easily, en masse and in diverse conditions. Carbon chemistry turns up elsewhere in the solar system. Saturn's largest moon, Titan, has methane clouds and hydrocarbon bays. Carbonaceous dust falls to Earth all the time—100 to 1,000 tons a day, although less than 1 percent arrives in a quantity large enough to analyze.[18] Meteorites carry libraries of amino acids foreign to life's repertoire of twenty. There are so many hydrocarbons and carbon-rich materials in the solar system that it's enough to convince many scientists that some of the Earth's petroleum and natural gas might have been produced by lifeless, geological forces—not by living things. Therefore not all are "fossil" fuels, pressure-cooked from once-living matter.[19]

University and government labs can manufacture prebiotic molecules under simulated conditions so easily it seems like kick-starting life must be as easy as falling off a log. Leslie Orgel, a prominent origin-of-life scientist at San Diego's Salk Institute, put it this way. "There are lots of ways of making organic compounds, so it's not really a question of, how on Earth could we ever have made any. It's rather a question of, we know of lots of ways you can make them. The real problem is to know which are the important ways that collaborated with each other, and we don't know that yet. Unfortunately, it happened 3.5 billion years ago, so it's hard to know which ones were the most important."[20]

To some extent it's challenging to say which ones are still important. Scientists have only begun to study the "deep carbon cycle," including questions about carbon in the Earth's lower crust, mantle, and core, what's there, how much, and what it does. A large, interdisciplinary team of scientists in the next couple of years will ask whether the Earth still has primordial carbon in its interior, how far down microbes permeate and affect the deeper movement of carbon, and "how do diamonds form, and what do diamonds reveal about sources and transport of deep carbon?"[21]

In the first half of the twentieth century, two scientists, Alexander Oparin in Russia and J. B. S. Haldane in the United Kingdom, independently honed their speculations on the origin of life by framing them within the principles of geochemistry. Each assumed that life must have emerged under an atmosphere of energy-rich gases. Their curiosity and extrapolation of observed geochemistry into informed conjecture laid the foundation for much future work.[22]

Haldane's and Oparin's work inspired the first experiments, which made worldwide headlines in 1953, when University of Chicago chemist Harold Urey and twenty-three-year-old graduate student Stanley Miller filled the first test-tube atmosphere. It is much more than a test tube, of course. Urey needed only an hour or so to design it. The university glassblower made it to the scientists' specifications. It is a rectangular tube, with thin glass piping leading from a bulbous "ocean" flask at the base up to a bulbous "atmosphere" complete with tungsten electrodes to simulate lightning. Tubing runs from the atmosphere back down to the ocean to complete the cycle. An original has long sat in a box most of the time in Sverdrup Hall at the Scripps Institution of Oceanography. On my visit, about a year and a half before Miller's passing in 2007, the apparatus was filled with trace amounts not of goop, but coffee, which professors use as a goop stand-in during undergraduate lectures. The Smithsonian Institution claimed a twin as a piece of history.

Miller sterilized the apparatus and filled it with methane (CH_4), hydrogen, ammonia (NH_3), and water. The earliest atmospheres lacked oxygen for reasons that take up most of the next chapter. Electrical sparking vaulted atoms out of their starter molecules and into new chemicals. Within a few days, a gooey black substance lined the tubing. Much of it was just that—black goo—but mixed in were some of life's likely materials, including four of the twenty amino acids that cells use to build proteins. Later experiments produced thirty-three amino acids, half of which life uses in proteins. The amino acids were formed from a flood of intermediates, such as hydrogen cyanide (HCN), formaldehyde, and acetone—which today is better known as nail polish remover. According to the primeval soup hypothesis, life may have evolved from an ocean of material so poisonous to animal life that half a whiff would have killed us.[23]

The Miller-Urey experiment has been repeated thousands of times with different gas combinations. The original study came under criticism as years ticked by, and scientists became convinced that the Earth's first atmosphere could not have been so enriched in the reactive gases—methane, ammonia, and hydrogen. Two scientists from the Scripps Institution of Oceanography revisited the experiment using carbon dioxide and nitrogen, a combination that failed to replicate the amino acids when Miller tried in 1983. Jeffrey Bada, a senior professor, and Jim Cleaves, then a research chemist, both colleagues and friends of Miller, realized the early Earth would have had more to work with than just those two relatively inert gases. Iron and carbonate minerals might have been present to cancel out the chemicals that interfere with amino acid synthesis. When Cleaves and Bada added those to the mix, the amino acids returned.[24]

Origins researchers have long posited that the atmosphere needed to do more than warm the Earth. It needed to provide energy-rich carbon and nitrogen gases that could react with each other and slime the Earth's shallow waters and rocky shores with carbonaceous film. The Sun's youth creates what geochemists cheekily refer to as the "Chinese restaurant paradox" (because "dim Sun" and "faint, young Sun" sound like garbled Cantonese phrases to the anglophone ear). The dim-Sun paradox is an attempt to reconcile the need for both a strong greenhouse and an atmosphere stocked with energy-rich molecules. Methane and ammonia are high-energy molecules, but the Sun's ultraviolet rays degrade them quickly. Hydrogen is also highly reactive, but the gas is light. Hydrogen balloons make helium balloons look like they are in no hurry to fly off anywhere. These gases would have been friendly to the emergence of life, but easily destroyed by ultraviolet light. Carbon dioxide is a powerful greenhouse gas and would have taken care of the weak-Sun problem, but may contain too little energy to jump-start life. Therein lay the problem.

All these gases may plausibly have been present. Methane might have made the skies look pink.[25] If the volcanoes continuously pumped methane into the air faster than the Sun destroyed it, a net accumulation could have helped build the feedstock of carbon molecules that inched chemistry closer to life. An atmosphere rich in carbon dioxide and methane, perhaps 1,000 ppm (parts per million) of each, addresses the dim-Sun problem and stocks a prebiotic soup with nutrients. An organic haze, not unlike that found on Saturn's moon Titan, might have enshrouded the early Earth, a hypothesis that has had some experimental backup.[26]

Evolution shaped prebiotic chemistry before life fully arrived. A. Graham Cairns-Smith of the University of Glasgow has intrigued (and infuriated) many scientists and nonscientists by speculating that early evolution may have taken place among inorganic minerals. He likens this dynamic to the way preindustrial societies built stone arches. They built a mound of dirt, assembled the stones atop it, and then removed the dirt, leaving a sturdy arch. In Cairns-Smith's mineral scenario, carbon compounds hitch themselves to a mineral replicator and assemble their own organic replicator from it. Eventually a "genetic takeover" separated the carbon replicator—a genetic code—from the mineral evolution, leaving only carbon-based life. This particular idea is difficult to test, which leaves it creative, reasoned conjecture. But minerals' potential role in the origin of life drives many important research programs today.

Universalities of life provide a more general frame for thinking about how evolution may have forged life from chemical precursors. All organisms share the same genetic template, the same core metabolism and molecular boundaries separating their internal chemistry from their environment—the prerequisites for life that Szathmáry abbreviates as template (T), metabolism (M), and boundary (B).

The template, or genetic code offers evidence for the uniformity of all life. At the core of the life sciences is the so-called Central Dogma of molecular biology. Every living thing stores genetic information in deoxyribonucleic acid (DNA), a macromolecule of two lovely parallel helices linked by those carbon-heavy nitrogenous bases, the "letters" of the genetic code. Five-carbon sugars alternate with phosphate molecules all the way down each helix. DNA's twin backbones curl because of the way carbon atoms in the sugar plug into oxygen atoms in the phosphate.

DNA is the storage medium that the cellular apparatus "reads" to build a cell. Ribonucleic acid (RNA) molecules come in limitless shapes and sizes, each with specific functions that are instrumental to the growth and regulation of cells. It is usually a single helix. Messenger RNA (mRNA) transcribes DNA's message and carries it to the cell's ribosome—itself essentially an RNA machine. There, transfer RNA (tRNA) takes the genetic message and aligns amino acids into a specific sequence. They fold into a protein, whose shape allows it to perform a specific function in the cell. The "groove" or "pit" in a protein enzyme has the perfect electrochemical shape to fit its substrate, which can be a hormone or another small molecule, which biochemists mercifully call "small molecules." Together they perform a specific action in the cell, like a molecular assembly line.

The Central Dogma of molecular biology exemplifies a fundamental principle of living matter (and inanimate matter). Chemistry seems like a sloppy mess to our senses, but it thrives on atomic precision; a molecule's structure is coincident with its function. Life is an electrochemical jigsaw puzzle, every piece dependent on physical contact with others for the system to work. Information, action, and structure are inseparable, in a manner reminiscent of the William Butler Yeats poem "Among School Children," which ends:

> O chestnut-tree, great-rooted blossomer,
> Are you the leaf, the blossom or the bole?
> O body swayed to music, O brightening glance,
> How can we know the dancer from the dance?

For biomolecules, there is no separating the dancer from the dance. DNA is life's parts list, written in a three-dimensional interlocking carbonaceous alphabet, encoding four information bits: adenine (A), guanine (G), thymine (T), and cytosine (C). Adenine and thymine pair exclusively with each other, as do guanine and cytosine. (In RNA, thymine [T] is replaced by uracil [U].) Together, each pair forms one of the rungs, spaced 0.34nm (nanometers) apart, all the way up the DNA ladder—2m (meters) of rungs crammed in every human cellular nucleus!

Atoms and molecules are, of course, very small, and many scientists and writers have their own pet metaphors. If a carbon-12 nucleus is the size of a soccer ball, sitting, say, in the gilded hand of the Prometheus sculpture above the Rockefeller Center ice skating rink in central Manhattan, the electrons are skipping about in a cloud whose diameter is about the length of Manhattan, ten miles. This tininess is complemented by the incredible speed of activity on the molecular level. Cells continuously unzip, unfurl, read, replicate, twist, and repack DNA at incredible speed. You have added about 17 million billion new base pairs (bp) in less time than it is taking to read this sentence—one second. The world of carbon moves quickly.

The "words" composed from the nucleic acid alphabet are three "letters" long. These sequences of three nucleotide base pairs (A, T/U, G or C) are called codons. Each codon is an order for one amino acid, the building blocks of proteins that have carbon at their cores. The codons—basically three rings of carbon and nitrogen, with attendant hydrogen and oxygen, all working together—yank the correct amino acid into place in a growing polyamide chain that will become a protein. The universality of these codons is further evidence that all life is related. There are only twenty amino acids, but sixty-four ways that the letters A, T/U, G, and C can be arranged in groups of three. That means some amino acids have more than one recipe in the genetic code. For example, RNA orders the amino acid serine with any of six codons (UCU, UCC, UCA, UCG, AGU, and AGC)—the most of any others. The vast, vast majority, if not all life, has the exact same mapping of three DNA and RNA nucleotide bases to the desired amino acid. As far as anyone knows, there's no reason for life to have selected the codons it has for life's twenty amino acids. In other words, evolution might have found different DNA and RNA sequences to code for different amino acids. The fact that they are identical is good evidence of a universal origin of life. And the language of biology—DNA, RNA, bases, codons, amino acids—saves everyone from having to describe everything explicitly in terms of carbon and other atoms of life.

The genome is too complicated for Earth chemistry to have solved the same gargantuan problem with the identical and enormously complex answer more than once. Consider the few, yet nonetheless unexpected, similarities between *Homo sapiens* and *Aquifex aeolicus*, (*A. aeolicus*) a 5 μm (micrometer) wide bacterium that lives in waters up to 95°C—nearly boiling. The *A. aeolicus* genome contains 1,551,335 positions for the nucleotide base pairs, A, T, G, or C.[27] There are four possible letters, the base pairs, that can each fill one of the 1,551,335 spots on the genome. That means the theoretical limit of unique *A. aeolicus* code combinations is $4^{1,551,335}$. That's four multiplied by itself 1,551,335 times, an inconceivably astronomical number, yet one that shrinks to a rounding error in comparison with the human genome. People carry around in each of their cellular nuclei about 3 billion bp of DNA. That makes the theoretical maximum number of sequences of the length of the human genome $4^{3,000,000,000}$.[28] As a reference, the number of all electrons in the Universe is estimated to be "just" 10^{80} or 4^{133}.

These numbers are so preposterously large that the likelihood for random overlap is all but mathematically zero. So if human beings and *A. aeolicus* have any genes—just one—in common, they are not random, and humans and *A. aeolicus* share a genetic code descended from a single ancestor. We share several dozen genes.[29]

The Central Dogma emerged during the 1960s, after Francis Crick and James Watson assembled their ball-and-stick model of the DNA molecule. Molecular biologists understood the role of RNA in expressing genes into proteins. Crick himself once called the Central Dogma "an idea for which there was no reasonable evidence."[30] The evidence came shortly thereafter, and keeps coming. Some theorists introduced another idea with "no reasonable evidence"—how this elegant system of complex structures might have operated at the beginning of life. Wouldn't it be simple, or at least simpler, if one supermolecule performed the information-storage role of DNA, the translation job of RNA, and the myriad functions of proteins? One molecule to carry the code, transmit it, and catalyze reactions—including its own creation.

Discoveries breathed new life into the idea fifteen years later, when University of Colorado scientist Thomas Cech discovered to everyone's astonishment that RNA splicing, previously attributed to proteins, can occur with no proteins in sight. RNA in certain circumstances splices itself. It doesn't always need proteins (though most RNA messages are spliced with the help of proteins). Sidney Altman of Yale University independently discovered this catalyzing property of RNA, and he and Cech shared the 1989 Nobel Prize in Chemistry. This discovery gave some evidence to support what earlier had merely been reasoned conjecture. Perhaps early cells needed only RNA to perform the work that DNA and proteins do with greater specificity and robustness today. Writing in the journal *Nature*, the Harvard molecular biologist Walter Gilbert dubbed this hypothetical predecessor to the Central Dogma the "RNA world."[31] In 2000, molecular biologists at Yale conducted x-ray studies on the ribosome, the cell's protein assembly center.[32] They found that decades of thought on life's protein factories missed the mark. The ribosome is not an edifice of protein supported by some RNA. It is an RNA machine bolted into place by a few proteins. Whether a world of RNA cells actually existed as viable life forms is likely an eternally theoretical question. This isn't: to a remarkable extent, we still live in an RNA world.

Metabolism also betrays a one-time emergence of life, according to another hypothesis. The notion of a "metabolism-first" origin of life is much younger than the primeval soup hypothesis first tested by Stanley Miller, and has the added challenge of being very difficult to study experimentally. Among origins researchers who try to build primitive, living chemical systems "metabolism seems to be the stepchild in the family," Szathmáry writes. What follows is therefore theoretical, but well constrained within well-established Earth and living chemistry.

All life retains a core chemical cycle. Organisms use energy and nutrients to build all the main kinds of biomolecules: the lipids (fatty acids) that form membranes, the sugars that store energy, amino acids, and consequently the nucleotide bases that are built from them. Microbes and plants are called autotrophs, or self-feeders, if they build their own food from nutrients and energy without other organic inputs. And all autotrophs in all ecosystems share a core cycle that builds life's library of fewer than five hundred molecular building blocks. All these molecules are smaller than 400 Da (daltons)—a standard measure of atomic size equal to one twelfth the mass of a carbon atom. Harold Morowitz is Robinson Professor of Biology and Natural Philosophy at George Mason University and an external faculty member of the Santa Fe Institute. He describes this universal metabolism as a "virtual fossil," and argues it is the most logical trace of how life first self-organized.

Morowitz, and Eric Smith of the Santa Fe Institute invert the common approach to how organisms react with their environment. It's no surprise that living things consume energy and organic materials. But that's a very life-centric way of thinking about the problem. In searching for life's origins, they argue, few seem to take the Earth's needs explicitly into account. What if life is the result of an early Earth bathed in energy with limited ways to degrade it into heat and radiate it back out to space? For analogies, they look elsewhere in nature, where energy channels "relax" disparities in conditions, such as air pressure or electrical charge. A lightning bolt channels electricity from the upper atmosphere to the ground. Temperature and pressure gradients create a channel that water and air fill by swirling into a hurricane. Perhaps life also began by exploiting differences in energy, creating a channel between, say, carbon dioxide—the carbon compound with the lowest energy—and methane (CH_4), which has the highest energy. Life could be the Earth's way of restoring equilibrium between high- and low-energy molecules.

The Earth's potential need for a biochemical steam valve may explain why life has persisted for 4 billion years, even as species die and extinctions periodically wipe out entire ecosystems. Life always bounces back, perhaps not because "it's a fighter" intrinsically but because the Earth needs it here to process energy. Smith and colleagues have estimated that atoms in light-absorbing photosynthetic material (which are mostly carbon, hydrogen, and oxygen) convert high-energy, visible light into heat energy a billion times more efficiently than the same atoms would in a nonbiological state.[33] Living matter relaxes the disparity between incoming solar energy and how well the Earth can degrade it into heat and send it back out to space.

Morowitz looks at the heat-loving, minimalist bacterium *A. aeolicus* as a living organism that contains the entire core metabolism. It has about 1,500 genes, perhaps two to four times the number needed in the smallest possible genome. His goal is to find a bright-line definition between chemistry and biology, where none now exists. "Is [*A. aeolicus*] predictable from first principles of chemistry? I believe that it will be. We haven't done it yet. I believe that it will be predictable from first principles of chemistry . . . How long have I been thinking about this? Fifty years? I have no idea how close we are. A good friend of mine in physics says a good idea takes 10 years and 15 minutes."

The third element in this conceptual breakdown of what proto-organisms needed, according to Szathmáry, is a boundary, or some predecessor to a cell membrane. Here, carbon chemistry has been kindest to experimentalists. Primitive bags of ubiquitous carbon molecules assemble by themselves. The reason this is so takes us to the core of why carbon molecules and water are such a good combination to make life.

Each of the three necessities for life—the code, the metabolism, and the boundary—is a function of carbon and its ability to bond, unbond, and rebond with the other atoms of life. Carbon makes strong bonds that structure tough molecules, such as fatty acids or sterols (think cholesterol). Just as important as its ability to hold molecules together is that carbon "knows" when to let bonds break so their atoms can recombine into new molecules. More reactive molecules typically include carbon-oxygen and carbon-nitrogen bonds. Carbon is more life's "Velcro" than, say, its "glue." Glue can be used only once. Velcro can be unfastened, re-fastened, and stay fastened as necessary.

Life being based on carbon is very different from *being* carbon. Diamond is crystalline carbon. So is graphite. Charcoal is the carbon of wood with the water cooked out of it. Fullerenes are molecular balls and tubes of carbon. Carbon nanofoam, a short-lived substance discovered only recently, even has magnetic properties at -183°C.[34] But none of these pure-carbon substances occurs in living things. Life demands that carbon atoms bond covalently, that they acquire a full complement of electrons through sharing—with the other so-called CHONPS atoms: carbon, hydrogen, oxygen, nitrogen, phosphorous, and sulfur. "Carbon-based life"—a science fiction cliché and real-world redundancy—also needs water as a solvent, nutrients from its environment, and a steady source of energy.

A cornerstone of living chemistry is the fact that electrically neutral molecules, alas, are not completely electrically neutral. This seemingly arcane detail of molecular behavior explains the interaction of water and carbon molecules, and by extension, a central characteristic of life. Some substances have a slight charge—called polarity—to them, a result of the internal arrangement of their positively charged nuclei and electrons "clouding" around it. Atoms attract their electrons with varying strength. This strength is measurable, and is called Electronegativity.

When two atoms with similar electronegativities bond, the molecule they form has little or no polar charge. The nuclei pull on electrons with about the same force, so the electrons don't favor one nucleus over another. Whether they make up your cellular membranes or your salad oil, hydrocarbons are basically nonpolar molecules because carbon and hydrogen tug on electrons with about the same force. They have minimal tendency to trend negative or positive.

That's not true of molecules built of atoms that draw electrons with varying strength. For example, the oxygen nucleus attracts electrons more strongly than hydrogen nuclei. When these atoms bond into water molecules (H_2O), they take on a boomerang-like shape. The oxygen vertex has a negative charge. The two hydrogen ends lean positive. Water is polar. These internal charges cause it to organize into short polymers, or molecular chains. The O of one molecule attracts an H of another; the O and Hs of that second molecule pull toward Hs and Os of third and fourth molecules, and so forth. It's like a tug of war, where the electron sits at the center of the rope. In carbon-hydrogen bonds, the electron doesn't move past that central starting line. In water molecules, the oxygen "team" pulls the electron farther in its direction than "team hydrogen" can. Oil and water don't mix because the hydrocarbons' nonpolarity and the water's polarity cause the molecules to separate as far away from each other as possible.

This information is critical to understanding the basic organization of living things. Hydrocarbons are stable under normal conditions, whether it's benzene in a gas tank or cholesterol in your arteries. You need heat or some other catalyst to speed the dissolution of C-C or C-H bonds. This situation changes if oxygen or nitrogen is introduced into the molecule. The mismatched electronegativities make these bonds more reactive, the electrons more likely to move. Animal and vegetable fats have reaction sites where carbon and oxygen share two pairs of electrons—a double bond. Burning fats or carbohydrates is not a metaphor. That's what's really going on in digestion. These carbon-oxygen double bonds are the sites where a cell begins to dismantle the fuel molecule and retrieve its energy. That makes combustion possible at our familiar 98.6°F. Nearly every reaction pathway in the universal metabolism of life takes advantage of the potential for change where a carbon atom is doubly bonded to a nitrogen or oxygen atom.[35]

Cellular boundaries or membranes unify the water-loving nature of phosphates and the water-hating nature of hydrocarbons into the same molecules. Experiments have demonstrated the ease with which cell-like membranes form spontaneously, almost like soap bubbles in the sink.[36] They do so for the same reasons that oil and water don't mix. Cellular membranes are commonly made up of a class of molecules called phospholipids. They have properties of both hydrocarbons *and* water-loving polar substances. When phospholipids reach a certain concentration threshold in water they self-organize into globules. The phosphate ends dissolve in water. The hydrocarbon ends drive away from the water, toward each other. Past another concentration threshold, the layers spontaneously collapse into two-molecule-thick sheets, then finally into hollow balls, or vesicles. This creates a macromolecule with key properties. The interior and exterior dissolve in water, yet a hydrocarbon barrier limits passage in and out.

In 1989, David Deamer tested these ideas on about 3 ounces of a meteorite given to his University of California–Santa Cruz lab by Chicago's Field Museum. The sample came from a 1969 meteorite that fell near Murchison, Australia, a goldmine in astrobiology research for its dumbfounding variety and volume of amino acids and other carbon molecules. The rock contained a bevy of organic material, including seventy different amino acids, most of which life doesn't use. Deamer cracked open the sample with minimal effort and separated out small white pebbles called chondrules, which typically appear in carbon-rich meteorites. He next applied fluorescent light, which revealed the presence of molecules ubiquitous in nature, conjoined rings of six carbon atoms. Indeed they showed up when Deamer applied the light. He isolated these hydrocarbons from the meteorite specks. He dried them on a glass microscope slide, added a water-based solution, and sure enough the material formed vesicles. This result demonstrated the ease with which it might have been possible for amphiphilic molecules flown in on meteors to survive impact, and self-organize, perhaps in tidal flats where drying and dousing occur regularly. Vesicles form so readily that the lab later showed how they might encapsulate genetic molecules or proteins. Two sheets of phospholipids sandwiched around a protein or RNA snippet under dry conditions, when rewetted, merge, enclosing the macromolecule.[37]

Vesicles address the boundary problem. Next is how prebiotic molecules were "selected, concentrated and organized" from starter chemicals and ended up inside phospholipid sacs.[38] For answers to these questions some scientists look closely at the surface of minerals and how they attract and bind certain kinds of organic molecules, aiming to build a ship in that bottle.

Minerals may have had a pivotal role in bonding molecules into polymers. The British scientist and thinker John Desmond Bernal is credited with suggesting a pivotal role for clays as sorters and binders of organic molecules in the late 1940s. But it took another half century for lab experiments to begin to mature. Unlike the tubes and bulbs of laboratory glass, clays have uncommon molecular surfaces. A handful of clay might not look like much, but it offers about 1,000 square feet of area to catalyze reactions.[39] Sheets of clay stack together like a "deck of cards," the layers glued together by powerful atomic forces.[40]

A particular kind of clay has drawn interest as a catalyst, for its presence on Earth, plausibly on early Earth, and even on Mars. Montmorillonite forms as volcanic ash settles and is worked over by weathering and other forces. Clay formations as thick as 16 meters underlie parts of the western United States. Industry mines it for use in everything from industrial abrasives to cat litter.[41] Catalysts play a central role in carbon science, from minerals in origins research to enzymes inside living cells to industrial technologies. Carbon is the foot soldier of the elements, capable of executing a boundless variety of molecular formations, but like an infantryman, takes orders for when to stand and when to invade.

Scientists have had success inducing RNA nucleotides to stick to montmorillonite. James Ferris of Rensselaer Polytechnic Institute has built on montmorillonite short RNA strings, up to fifty units long in one day. Ferris's research supports the idea that the origin of life required inorganic catalysts to bind components. Many issues remain, including probable ways that A, C, G, and U might have bonded to ribose and phosphates to make RNA nucleotides in the first place. But this research supports the RNA world hypothesis. "I may be wrong in assuming the RNA world arose from pre-biotic reactions on the primitive Earth," Ferris has written. "But I am convinced that mineral and metal-ion catalysis was absolutely essential for the formation of the complex organic structures that were necessary for the first life on Earth."[42]

RNA molecules can grow by sticking to clays. They also stick to many other important substances that some scientists think must have been necessary for life to have begun in an "RNA world." Two properties of RNA underpin this hypothesis: RNA's abilities to keep a genetic code, like DNA, and to catalyze reactions, like proteins. A powerful experimental tool has added more credibility to the RNA world hypothesis by demonstrating RNA molecules' tendency to mutate, acquire new properties, and save desirable ones for further development. If this sounds like evolution, that's because it is. In vitro evolution, and its sibling directed evolution, has already proved an essential investigative tool to explore the chemical possibilities that may have allowed increasingly complex carbon-rich molecules to find a stable genetic system and the metabolism to keep it going.

The first cells would have needed a way to incorporate nucleotides, amino acids, and life's myriad small molecules. Scientists have shown that RNAs can evolve to stick to these things and much more—viruses, tissues, ions. In addition to this ability to grab key molecules, RNA can speed reactions along, including the linking of two carbon atoms, in a way similar to one of chemistry's most ubiquitous reactions. This work has grown up in parallel to the biochemists' gradual realization that RNA molecules play a much larger role in how cells work than they are historically credited for, even challenging the Central Dogma of molecular biology as a complete central dogma.

The variety of RNA molecules is a function of chemical performance a level above carbon's atomic versatility. Like DNA, RNA can store practically infinite potential coding sequences. But unlike DNA, RNA is usually a single strand, not a double helix. This structural feature gives it flexibility, literally, to take on practically infinite shapes to complement its coding prowess. In 1990, Jack Szostak's lab at Harvard University demonstrated a way to harness this molecular diversity. In this experiment, they chose organic dyes—rings of carbon with distinctive side groups—as the target molecules they hoped to find RNA molecules to stick to. Szostak and Andrew Ellington ran a soup of 10^{13} random RNA sequences through laboratory glassware. A tiny fraction stuck to the dye molecules. They separated these successful matches and generated a new library of random RNAs similar to them using a standard laboratory technique. They ran ever-more refined RNA concoctions through the apparatus, and in so doing grew a library of RNAs that highly evolved to fit the targets. They called the evolved RNAs "aptamers," from the Latin *aptus* for "to fit."

This early work led directly to one of the most ambitious origin-of-life programs in the field. Given the tendency of phospholipid vesicles to self-assemble and of RNA to code, catalyze, and select precise targets, Szostak's lab has embedded functioning RNAs within vesicles, as a potential segue toward the first synthetic organisms.

Szostak's lab has achieved many noteworthy successes involving RNA-filled vesicles. Vesicles can grow and divide on their own, driven by osmotic pressures. The lab has shown that a simple RNA machine can work inside a vesicle. These tiny systems can grow larger by pilfering fatty acids from nearby, empty vesicles—driven only by simple chemical forces. Vesicles "compete" for materials based only on the pressure and chemical gradients in the system, without complicated enzymes. Membrane growth even creates an energy pathway that proto-organisms might someday be able to use to transport amino acids and other nutrients into the vesicle. These experiments show that the microscopic fatty acid bags can perform what looks like "primitive forms of Darwinian competition and energy storage," suggesting a pathway from prebiotic macromolecules to replicating, adapting, living things.

Szostak and his colleagues strive to build a chemical system that develops its own novel properties—that evolves. For an organism to be considered alive, it needs more than a self-replicating RNA isolated inside a membrane compartment. The RNA and the membrane need to be able to interact in a way to bring novelty into the system. "For our work, evolution is the goal," Szostak says. "What we're doing has come to be a project in chemistry, and trying to get evolutionary behavior out of chemistry."[43]

Achievement of this goal would provide evidence not for how life actually evolved, but how chemistry behaves, and might have behaved on the early Earth. Four billion years separates science from the actual origin of life. Since Stanley Miller's first primeval soup experiments in 1953, origin-of-life research has strived to understand plausible chemical conditions that contributed to life's origin.

Simple evolution in RNA-filled vesicles will be a major accomplishment in origins research. It won't show how life evolved originally, only that certain pathways seem to enable it.

Science knows much more about how evolution works than how to start it from scratch. Everything from dog breeding to genetically modified crop production to molecular biology works on already-thriving evolutionary systems.

The definition of evolution grows more complicated as physicists and chemists peek over the shoulders of biologists and as the latter probe how cells build themselves from their genetic codes. Thumbnail definitions 150 years old still suffice most of the time. Life distinguishes itself from other phenomena that appear to grow, such as crystals, because it reproduces imperfectly and consumes and spends energy working to stay alive through the sieve of natural selection. Descent with modification generates novelty into the genome and phenotype, or an organism's collection of observable traits, some of which will be selected and others will not. Five points are worth keeping in mind as a description of evolution, summarized handily by Nicholas H. Barton and colleagues in the book *Evolution*, points that we'll revisit periodically in the next four chapters:

- Evolution occurs through a universal cellular apparatus—RNAs, ribosomes, proteins, small molecules, and environmental inputs and influences—that cobbles DNA's message into the phenotype. The earliest life may not have required all of these moving parts. In vitro and directed evolution demonstrate how evolution can occur within populations of RNA molecules alone. But the evolution of "life as we know it" requires a complex apparatus.

- Biologists distinguish between this universal cellular apparatus and that belonging to any particular organism of any particular species. Only populations evolve; individuals adapt. Living things evolve as individuals' selection and reproduction reshapes the gene pool.

- Since before Darwin published *Origin of Species,* in 1859, natural philosophers thought of life as a branching tree. Molecular biology has challenged this analogy. In the microbial world, branches cross, merging and diverging. Microbial organisms can gain or lose genes while they are still alive. This horizontal gene transfer makes evolution "look" more like a dense bush than a tree.

- Evolution has no goal. Earth history has not been a march to humanity any more than it has been to howler monkeys, sharks, *A. aeolicus,* bacteria that cause tooth decay, or HIV.

- Finally, evolution's engine is natural selection among random mutations. An organism's inherent traits, its predators, its environment, and many other things can impose selection—whether its genes carry into the next generation.[44]

Apropos to the global story of carbon, evolution is the dynamic by which life expands and contracts its appetite for carbon, being transformed by the Earth system and transforming it in the process.

3 | The Flood

Molecular Fossils and the Great Greenhouse Collapse

For we are living
In a bacterial world
And I am a bacterial girl.

—Suzy Wagner (with apologies to Madonna)

Very small things evolved from other very small things more than 2.7 billion years ago. That's not surprising. Microbes mutate and swap genes all the time. That's how bacteria become immune to antibiotics. What's surprising is that these very small things, called cyanobacteria, thrived in such numbers that their waste destabilized the Earth's ancient carbon cycle over several hundred million years. Scientists have long tried to sketch out the early evolution of these tiny organisms, with novel insights in recent years.

Of the untold millions of species that have lived and died since life emerged, cyanobacteria stand out as organisms of superlative influence. They invented photosynthesis as we know it—making carbohydrates with sunlight, CO_2, and water, while emitting molecular oxygen (O_2) as waste. Sometimes called "blue-green algae," they are no more algae than seahorses are horses. Other kinds of photosynthesis existed before then, and still do, but none used water as a source of electrons and protons needed to help cook the carbon dioxide into sugar, and none pumped out oxygen in the process.

The story of cyanobacteria is demonstrative of how evolution could hit upon successful innovations whose aggregated effect cascaded through the Earth system and reshaped the carbon cycle. One way to think about evolution's impact on the Earth system is as a nimble, shape-shifting carbon mover. In cyanobacteria, evolutionary forces hit upon possibly the most influential innovation since the beginning of life—the ability to make carbohydrates from sunlight, airborne carbon, and water. This innovation not only led many, many ages later to the rise of red algae and green plants. But the novel skill of cyanobacteria—the exchange of atmospheric carbon dioxide for atmospheric oxygen—threw the carbon cycle into catastrophic disarray and then reinvention.

Some scientists have argued that the rise of atmospheric oxygen killed off many species of microbes, but it's hard to demonstrate. Rocks that old offer precious few fossils and little tangible evidence of what microbes may have disappeared when the wheels came off the carbon cycle. What's certain is that many microbes accustomed to living in an atmosphere of nitrogen, methane, and carbon dioxide suffocate in oxygen. Species that survived the speculated "oxygen holocaust"[1] receded to dark corners of the Earth, where the gas does not penetrate. Another argument holds that oxygen more likely triggered an expansion of the biosphere into previously sterile waters, where organisms could continue to thrive in the absence of oxygen.[2] Anaerobic cells still flourish in places oxygen can't reach—intestines, bogs, the mud at the bottom of the sea, in rock two miles below the surface. Today's atmosphere is about 78 percent nitrogen, 0.9 percent argon, 0.1 percent water vapor and trace gases, and 21 percent oxygen—the most abundant, transformative pollutant the Earth has ever coped with.

The story of carbon is also oxygen's tale. They are the Lennon and McCartney of the periodic table. They each have solo careers, but neither is as compelling as their ensemble work. The periodic table of the elements is a powerful demonstration that the Universe is ordered and that the scientific mind can discern it. Dmitri Mendeleev, who standardized the recipe for vodka and advised the tsar about the nascent global oil industry, assembled the table in 1869 based on the properties of the known elements. His genius lay in something even more powerful than pattern recognition. He predicted that scientists would find elements missing from his table and foretold their properties. In 1871, germanium came along just as he had said. Two others followed, and Mendeleev's fame was assured. Perhaps the graphical monotony of the periodic table obscures the order of the Universe it unveils.

Before cyanobacteria evolved, no known force in the Universe existed specifically to split water molecules into hydrogen ions (protons), electrons, and oxygen.[3] Researchers disagree about when the cyanobacteria showed up. Blinded by the passage of time, scientists who study the deep past reconstruct past climates from sedimentary rocks and their isotope patterns. The cyanobacteria may have left traceable carbon behind them 2.7 billion years ago—as long ago as there are seconds in the life of a centenarian—or 2.3 billion years ago. A persistent but marginal argument puts them in Australia 3.8 billion years ago. Biomarkers, or molecular fossils, more than 3.5 billion years old may once have belonged to cyanobacteria, though that date has come under scrutiny and wears a Roger Maris-like asterisk. The deeper scientists look into the past the harder it is to focus. What matters most in science aren't factoids, but the shifting, ever-refined, always-incomplete assemblage of evidence, submitted by professionals using their best judgment, for evaluation by other professionals using their best judgment. In science, the means justify the ends.

The Earth is for practical purposes a closed material system. The amount of carbon, water, and other material in the Earth system doesn't change. Matter is conserved, to use the scientific word. Nothing leaves, so it just cycles from air to land to sea to sediment to mantle and back again. The law of conservation of matter frames the observation that tiny living things can alter the entire Earth. Little things, en masse and however defined—atoms, molecules, cyanobacteria, shelled algae, trees, cars—can cut pathways for carbon that over time change its global flow, and by extension, the conditions for life on Earth. Common sense resists the notion that any number of organisms too small to see, over any period of time, can change a planet too large to embrace. Cyanobacteria's disruption of the carbon cycle seems less plausible than the million monkeys eventually typing Shakespeare. But given the incomprehensible span of Earth history, gazillions of bacterial species producing umpteen gazillion individuals a day makes Shakespeare look like inevitability. Molecule by molecule, cell by cell, year by year, for as long as hundreds of millions of years, cyanobacteria pushed out enough oxygen to dramatically upset an atmosphere dominated by methane, carbon dioxide, and some ethane, a two-carbon gas.[4] Somehow common sense falls apart upon examination of common events.

Peer-reviewed journal articles are scientific professionals' traditional way of saying, "Hey! Look what we found!" The authors' colleagues look at what they found, and decide for themselves whether it's interesting, whether the results are replicable, and if so, how they fit into the big picture. The American judicial system provides a useful comparison, to an extent. The highest authority we can turn to for adjudication of alleged criminal wrongdoing is fellow citizens, guided by a professional judge, reaching a conclusion based on physical evidence. Similarly, the highest authority in science is the professional community itself, mediated historically by the editors of peer-reviewed journals. "A scientific statement is an act of communication," says Jan Fridthjof Bernt of the Norwegian Academy of Sciences.[5] Knowledge emerges only in dialogue and only with evidence. If a scientist doesn't find flaws in his or her own work, someone else very well may. Science is structured, ideally, to keep everything transparent and everybody honest.

Cyanobacteria belong to the phylum cyanophyta, which includes more than two thousand species. Even today they live almost everywhere sunlight touches water.[6] With an estimated population of about 10^{27}, you don't need to look far to see them at all. The prefix "cyano" might look familiar if you have ever replaced a color printer's ink cartridges. They usually come in three colors, yellow, magenta, and cyan, or blue. The bacteria are really more cyan-*ish* than blue. (This "cyano" prefix means blue; the same prefix in cyanogen or cyanide means a molecule that has a carbon-nitrogen group in it.) Colors vary given the array of light-catching pigments a cell inherits. Some cyanobacteria wax green, some red. Chameleon-like species can change their colors depending on the amount of light they receive from each point on the spectrum of white light—red, orange, yellow, green, blue, indigo, violet. One species, *Prochlorococcus*, is thought to be the most populous single species on Earth. It and *Synechococcus* fix more carbon into living matter than any other species.[7]

If you develop a mild rash after splashing around in the ocean, cyanobacteria may have lodged in your skin and caused "swimmer's itch." They live on rocks, in soil, and on the underside of tree leaves in tropical forests. Good luck seeding a desert oasis without cyanobacteria soaking in some starter puddle.[8] About 30 percent of the oxygen you inhale was manufactured by them. They are responsible indirectly for the other 70 percent, too. Plants and trees owe their carbon-fixing skills to these water-splitting microworkers. Cyanobacteria stick to building exteriors in stone ecosystems that give a new twist to the term wallflower.[9] The next time you go downtown, no matter where you live, look for the filthiest white building you can find. Those stone walls host microbial ecosystems that include evolution's revolutionaries, sitting unglamorously amid the filth.

For most of their history, cyanobacteria lived somewhere other than the cut-limestone exteriors of urban office buildings. The more old-fashioned among them still live like their pre-urban ancestors. "Other-timely" isn't a word, but perhaps it should be. It makes a sensible description of these almost entirely vanquished colonies. "Otherworldly" is technically incorrect, but has the right feeling. Microbial communities called stromatolites still thrive in a few otherworldly ocean bays. Usually shallow, salt rich, and continuously drowned and dried by tides, these sites are your best bet to see what the Earth's shores must have looked like long before any organism had eyes to see them. Half-meter-tall, ele-phantine limestone stumps grow off Lee Stocking Island in the Ba-hamas's Exuma Cays. Living stromatolites have even been found in freshwater, such as Lake Vermillon, in northern Minnesota.[10] But Aus-tralia is many paleobiologists' favorite playground.

The westernmost part of Australia offers scientists a rich mine of both living fossils and fossilized life. The whole area resides on a landmass called the Pilbara Craton, an iron-rich flatland that rock-recycling forces have left alone for 3.5 billion years. The most solid current biomarker evidence may be well-preserved 3.4-billion-year-old stromatolites from the Strelley Pool Chert in western Australia.[11] Bacteria may not have invented photosynthesis there, but this region offers its most ancient and newest traces. Australian scientists discovered 3,000-year-old stromatolites in 1954 at a site called Shark Bay.

In Shark Bay, microbial life proceeds as it has since cyanobacteria were the freaks of nature. More than fifty species of cyanobacteria thrive atop a 5 mm thick blanket of ooze. They are the stromatolites' top chefs, frying CO_2 into nutritious molecules by day and fixing nitrogen into ammonia by night. Beneath them in the mat, microbes live off each other's by-products in an ecosystem of squish. Billions of cells rely on each other for nutrition and protection. Sulfur-breathing microbes soak up the long wavelength photons that cyanobacteria don't catch, with which they transform others' waste into sulfuric acid and energy. Purple sulfur bacteria live off elemental sulfur, thiosulfate, and hydrogen sulfide. Fermenters break down sugar into alcohol and carbon dioxide.[12] (Their distant descendants have done well for themselves in the beer industry.) The stromatolite mounds grow because each generation leaves behind a thin layer of limestone, or calcium carbonate.

Cyanobacteria are an influential clade within the wider world of bacteria. In terms of species diversity, population, and biomass, we are living in a bacterial world. An estimated 5,000 billion billion billion (5×10^{30}) microbes use nearly as much carbon as all plants do.[13] About a million bacteria and more than 10 million viruses inhabit the average milliliter of seawater.[14] We belong to the oddball branch of the tree of life, that is, to the extent you can separate humans from bacteria, our most distant cousins. Each of us is an economy of differentiated complex cells crawling with microbial fellow travelers. Our cells are the result of several microbial mergers and acquisitions. Life as we know it is made up of cells, which themselves show traces of ancient organisms that moved in and never left.

We give bacteria a hard time, mostly because of a few bad species. Some species kill humans, such as *Yersinia pestis* (plague) or *Bacillus anthracis* (anthrax). Bacteria—the good, the bad, or the indifferent—generally coexist peacefully with humans, where we aren't codependent. When babies are born, they are slathered with bacteria as they move through the birth canal. Yet as soon as the cord is cut, doctors drop erythromycin on their eyes to prevent bacterial infection. Kenneth Todar of the University of Wisconsin–Madison has estimated that we humans have as many microbes in our guts as we have cells in our bodies. And that's not including the gazillion microbes on our skin and in our mouths.[15] They cover your keyboard, doorknobs, and, of course, exterior concrete. Bacteria fill so many niches and evolve so quickly, there is no estimate for the number of species to complement the 1.8 million or so species of plants and animals. It could be much more. It could be less, depending on how species are defined. Bacteria challenge species classification in general, and cyanobacteria are no exception.[16] "The absence of detailed knowledge of prokaryotic diversity is a major omission in our knowledge of life on earth," three professors from the University of Georgia–Athens wrote in

1998.[17] Two of the three domains of living things, unofficially called prokaryotes ("pro carry oats"[18]) are distinguished from eukaryotes ("you carry oats") by having "naked" DNA—no nucleus.

The roots of the tree of life look very different today than they did even twenty years ago. Three domains envelop all life: Bacteria, Archaea, and Eukaryota. Archaea are single-celled microbes, but differences in the way they express their genetic code distinguish them from bacteria. In many cases Archaea and Bacteria thrive in conditions that would kill life as we know it—at deep-ocean volcanic vents, beneath Arctic ice, in extremely salty bays or in caustic acids, in rock two miles below the Earth's surface. Their penchant for these environments gives the hardiest microbes the nickname "extremophiles." Eukaryotes make up life as we know it. Elephants, mushrooms, cacti, goldfish—everything alive other than bacteria or archaea—are eukaryotic cells, or federations of them. They store their DNA in membrane-bound nuclei. They also have little organs, called organelles, where cells retrieve energy from fuel or, in plants, fix atmospheric carbon into carbohydrates.

Scientists assume that the primeval ancestors of today's cyanobacteria turned CO_2 to carbohydrate in the same way that their modern descendants do. It's a reasonable assumption. "Assumption" has a more serious meaning in science than it does in everyday speech. An assumption is a statement about how the world works, based on observation, logic, or both, and whenever possible is tested for its viability. This assumption, that today's cyanobacteria live like their predecessors, is based on the observation that life tends to conserve the most essential evolutionary inventions across diverse species. Just as there is no reason to think that life began more than once, there is no evidence to suspect that organisms invented paths to oxygenic photosynthesis more than once. A one-time invention of photosynthesis and subsequent evolutionary dissemination of this identical, complex innovation down through bacteria, algae, plants, and trees is more trustworthy than the emergence of such a complicated system, from scratch, many times. Occam's razor is worth keeping in mind. The explanation that requires the fewest assumptions has the highest probability of being the correct one.

Scientists postulate that ancient organisms earned a living the same way their extant progeny do, a doctrine called "actualism," or "actualistic paleontology."[19] This is an important variation from the geological axiom stating that physical change on Earth—plate tectonics, volcanoes, rock weathering—has always worked in exactly the same way. Actualism means that unless contrary evidence turns up, the best guess is that an ancient organism made a living akin to its extant descendants. This doctrine is thought to yield the best possible explanation, given that it is often the only one. "Best" and "good" explanations in the Earth's deepest past are far from the same. Actualism lets scientists tour the path of carbon through photosynthesis, agnostic (but cognizant) of whether it is today or 2 billion years ago.

Cyanobacteria actually can tell time using a mechanism similar to our circadian clock, from the Latin meaning "about a day." Three billion years of sunrises and sunsets trained cyanobacteria to anticipate daybreak and nightfall. When life depends on maximizing sunlight, keeping track of when it appears and disappears becomes a compelling exercise. Dawn till dusk, cyanobacteria run microscopic factories, soldering hydrogen and carbon dioxide into carbohydrates. The circadian rhythm of cyanobacteria works surprisingly like a watch. Proteins and small molecules substitute for the quartz oscillator, gears, and springs strapped to the wrist. A change in temperature, humidity, or light triggers a protein oscillator, which turns a cycle of reactions that "winds up" the microbe for the day's work. The time of sunrises and sunsets changes gradually throughout the year, so cyanobacteria must continuously reset their clocks. Sometimes they fail to adjust to the Sun's increasingly early or late rises. Susan Golden of Texas A&M University playfully refers to this phenomenon as "jet lag."[20] It's the same basic principle. The circadian rhythm is interrupted and must adjust.

The Sun rises. Photons bathe Shark Bay with morning light. It is today, or long ago. Light particles careen toward a cyanobacterium bobbing a millimeter or so from the top of a stromatolite—the first catch of the day. Sun yields to Earth, photons to electrons, physics to chemistry, chemistry to life. The photons energize antennae-like molecules that absorb blue and red wavelengths within visible light. The receptors pass the photon's energy toward the cell, through the all-important chlorophyll molecule. Chlorophyll is the light-sensitive pigment that absorbs photons' energy and passes it into the cells of photosynthetic organisms. It's a lovely molecule, carbon's geometry forming a cloverleaf frame. Fifty-five carbons in all build the scaffolding, at the center of which sit four nitrogen atoms with a magnesium bull's eye. A long hydrocarbon chain tethers chlorophyll to the cell's photosynthetic membrane, which is studded with proteins that carry electrons forward.[21]

Everybody knows that plants need water. What happens to it once it's poured is less immediately clear. Plants—channeling a process original to cyanobacteria—break down water molecules (H_2O) as a way to prepare for incoming carbon. They have a special protein complex that dismantles water, removing two electrons and two protons, all of which end up in energy-rich molecules that help usher carbon into a cell. So for every two water molecules they mine, they produce one molecule of O_2 waste. That all land plants and algae have the same protein complexes—called photosystems I and II—further suggests the green plants and red algae are essentially ornate cyanobacterial condominiums.

The electrons removed from water zip through a photosynthetic membrane, ending up in a kind of battery molecule. Protons can't ride through the membrane, and get trapped behind it. Protons freed from water hang out behind the membrane, separated from the cyanobacteria's innards. They must rely on portals to move from one side of the membrane to another. They clamor toward it, but are limited in the quantity that can pass. The only door is straight ahead, through a merry-go-round-looking machine that leads to the cell interior. The buildup of protons is so powerful that it becomes an energy carrier—analogous to a hydroelectric dam—channeling energy that enables the cell to produce life's universal fuel molecule, ATP. Scientists call imbalances such as this a gradient, broadly analogous to the lightning and hurricanes described by Harold Morowitz in the last chapter. Those protons are stuck in a high-pressure system and want to get out.

The merry-go-round, a protein complex called ATP synthase, fits protons between a phosphate ion and a spent cellular battery called adenosine *di*phosphate. This union of ADP, protons, and phosphate ions makes ATP, or adenosine *tri*phosphate.[22] The cell's batteries are charged, waiting to unleash on the first carbon dioxide molecules that cross its path. That leads us to the big moment.

With the Sun's energy transformed and temporarily stored in molecular batteries, the cyanobacteria is primed to pluck carbon from the air. Carbon dioxide molecules hover picometers above a cyanobacterium. Air pressure and the draw of the cell push and pull the molecule inside, where two molecular handlers await within the cell system. One handler is a five-carbon sugar. The CO_2 molecule can't find this molecule on its own. It needs a specific enzyme, which is officially nicknamed Rubisco, to guide it into the right spot. Rubisco might be the most abundant enzyme on the planet and makes up about half the protein in leaves' chloroplasts.[23] These two molecules, the enzyme and its small-molecule substrate, incorporate carbon into a carbohydrate. When Rubisco fixes CO_2 into the substrate, the resulting structure is an unwieldy sugar that cracks into identical halves, each containing three carbon atoms. Some plants, such as corn, are called C4 plants because a four-carbon molecule, rather than a three-carbon one, is the first product of the reaction.[24] So proceed the "dark reactions" of photosynthesis. Cells assemble and dismantle chemical intermediaries until they make the stable six-carbon sugar, glucose.

Carbon's architectural nimbleness is so great that scientists need a special chemical language to describe what's going on. It's easy to lose track of carbon and companions very quickly in these many molecules; in a way the shared language of biology and chemistry might be thought of as the 33 million names of carbon, each structure distinguishable by its descriptive name. Carbohydrates all have the same general formula, CH_2O—a 1:2:1 ratio. But that formula doesn't tell you how many C's, H's, and O's, and it doesn't tell you how they are arranged. In fact, it's pretty useless. Dozens of carbohydrates have the general formula CH_2O. The dark reactions of photosynthesis produce $C_6H_{12}O_6$. That formula provides more information about the molecule but still doesn't explain the most important information. What is the shape? Six carbon, twelve hydrogen, and six oxygen can form any of sixteen possible arrangements—and that's not including sixteen mirror-image molecules, each "the same and not the same" as its chiral counterpart.[25] To specify which one, scientists need the precision of chemistrese—a language specific to chemistry, in which every syllable indicates the number, position, or type of a part of a molecule.

Tracing the path of carbon in photosynthesis from CO_2 to glucose took more than a decade of inventive, tedious research. University of California–Berkeley scientists figured it out by working with a radioactive isotope of carbon, still a recent discovery in the late 1940s.

Isotopes are keys to understanding how stars produce energy and how the Earth and life operate. Isotopes of carbon, oxygen, sulfur, uranium, lead, and other elements are indispensable clues to the past. They indicate the age of the planet and the meteorites that occasionally strike it; what kind of organisms lived where, how long ago; when the climate changed, and how sharply. Carbon-12 makes up nearly 99 percent of the Earth's carbon. Carbon-13 is also a stable nucleus, and composes just more than 1 percent. Living things tolerate this heavier isotope in varying amounts but generally much less than its natural occurrence on Earth. Why build a house with thirteen-pound bricks if they are otherwise no different from twelve-pound ones? By measuring the ratio of carbon isotopes in ancient rocks, for example, scientists can infer what kinds of organisms once lived there. If rocks have about a twenty parts per thousand lower quantity of carbon-13, something probably lived and died there. Variations in the carbon-13 ratio are powerful indicators of what kind of organism left its atoms behind. For example, methane-producing bacteria have an extremely low tolerance for carbon-13. So if organic rocks have very little of it, scientists might infer, considering other evidence, that methanogens lived there. Isotopes also help us to learn about the present. The same isotopic properties geologists seek out in carbonate rocks can reveal which elite cyclists and baseball players cheat by using banned performance-enhancing substances.

Radioactive isotopes have many uses in recent history. Carbon-14 is unstable—radioactive. Archeologists rely on it to date materials younger than 50,000 years. This isotope has eight neutrons in its nucleus, besides its six protons. It forms in the atmosphere, when cosmic rays hit nitrogen, but makes up a tiny fraction of the Earth's overall carbon. Over 5,730 years, half of a given quantity of the material will decay back into nitrogen. By measuring the ratio of radioactive carbon to nitrogen, archeologists, art historians, and geologists can put a fairly specific date on a material's origin, within a forty-year margin of error on either side. There's even carbon-11, which is highly unstable; half of it turns to boron in about twenty-one minutes.[26]

Melvin Calvin, Andrew Benson, and James Bassham fed algae with carbon dioxide known to contain abundant carbon-14, so that they could detect the carbon inside the organism. Experiments using the even more unstable carbon isotope, carbon-11, proved too difficult for previous researchers.

They ran the experiments many, many times, killing the algae at shorter and shorter intervals, and detecting the carbon-14 in the intermediary molecules, until they figured out the line of reactions carbon moves through before ending up in glucose. Calvin said this about how he and his colleagues pulled it off. "It's no trick to get the right answer when you have all the data. The real creative trick is to get the right answer when you have only half of the data in hand and half of it is wrong and you don't know which half."[27]

Since paleontologists cannot re-create in a lab experiment the conditions of life long gone, they look for clues instead. The most helpful clues from the deep past are biomarkers, molecular fossils, traces of chemicals made by organisms but degraded over time. As any detective-story aficionado knows, crime scenes go cold within hours. The same is true of dead ecosystems, on a much longer timescale. Unless exceptionally well preserved, DNA decays too quickly. Cyanobacterial fossils—their recalcitrant outer sheaths—stretch back some 2.1 billion years.[28] To see back farther than that, geochemists must look for the remains of molecules found in cyanobacteria.

The oldest geology goes back billions of years—one, two, three, four, *any* billion is too big a number to understand. Try even getting your head around just a million. Charles Darwin relied on a colleague's analogy. "Take a narrow strip of paper, eighty-three feet, four inches in length, and stretch it along the wall of a large hall; then mark off at one end the tenth of an inch. This tenth of an inch will represent one hundred years, and the entire strip a million years."[29] That's just a million years. A million years is to the history of Earth what six and a half days is to an eighty-year-old. Geologists function in deep time—the temporal equivalent of deep space. Even scientists can't really apprehend it. They just go by the numbers. As the environmental scientist William Ruddiman has written, "I suspect that even those of us who spend our entire careers studying aspects of Earth's history have no real grasp of their immensity."[30]

On average, rocks last about 250 million years before the mantle recalls and recasts them, or they are pushed up like the Himalayas and worn away. That means that half the rocks from 500 million years ago are gone, and 90 percent of rocks from the truly ancient past.[31] Rocks that somehow escaped tectonic reshaping are the only original documents we have about the planet's past. Scientists know, and fear, that some scientific treasures have economic value as well. Geologists must report the location of fossil finds in their scientific papers, but also shield it in a way so that their competitors—and fossil hunters—don't gut a terrain that sat uncontaminated for 3 billion years. At this writing, a 3.5-inch round, polished, stromatolite fossil purportedly 2.2 billion years old is selling for $69 on eBay, plus shipping. Others range in price by an order of magnitude in either direction.[32]

Scientists who work in the deep past are careful to distinguish "best current evidence" from "solid" or even "good" evidence. As much as scientists of the deep past work like detectives, they retain an element of the proverbial drunk who loses his keys late at night and only searches for them beneath street lamps because that's where the light is. The fossil record is incomplete and arbitrarily preserved. Scientists can only look for keys to the past "where the light is." Light shines brightest on the Earth's oldest mass graves, dark rocks high in organic carbon but not metamorphosized. Only a few places in the world are old enough to contain possible specks of the earliest life: southern Africa, Canada, possibly Greenland, and western Australia.

Marble Bar, which suffers as the most heat-scorched town in Australia, appears on most globes, odd for a town of four hundred. Spiders, scorpions, and snakes easily outnumber people. Travelers passing through the desert wilderness will drive one hundred miles out of their way, down arrow-straight iron flats, just to get a beer. Exiting Marble Bar, signs warn travelers: NEXT FUEL STOP 650 KM, KANGAROO CROSSING, NEXT 250 KM. Trains cause the ground to vibrate for dozens of miles in all directions. Roads unfurl straight, without deviation or exit. There's just nothing, unless you are a geologist.

Even in our high-tech age, geologists have only one way to acquire lab samples. They have to visit far-flung places like greater Marble Bar and chip at the Earth with hammers. Geology is an intimate science. Geologists crawl about our planetary body, rubbing and memorizing its distinguishing birthmarks and scars, listening for echoes of breath, licking rocks. They photograph fossils, usually with a pick, pen, pocketknife, or coin for scale. If you had no idea what was going on, you might wonder why so many people travel so far from home, to barren valleys, just to photograph their picks, pens, pocketknives, and coins. Separated from their surroundings for the first time in 3 billion years, the rocks are piled into boxes, boxes onto the truck.

Earth chemists have to know what to look for and where to find it. Organic carbon (once a part of living matter) is stored underground in limestone, sandstone, organic-rich black rocks, or gunk. Paleobiologists often make use of drill cores taken by mining or oil companies; biomarkers often help show the way to valuable resources in the extractive industries. Kerogen contains millions of kinds of molecules, morphed by pressure and heat into what on paper looks like mangled chicken wire. By heating the rocks back in the lab scientists can separate molecules by their mass and analyze what signatures may have been left in a given sample.

For a long time, heating rocks until they bled oil was considered an unproductive exercise. The oil was thought to be contaminated, made of geologically more recent hydrocarbons that seeped down to lower strata. Researchers were further discouraged by the assumption that biological molecules would degrade beyond recognition.

Intriguing work published in 1999 challenged this rule of thumb. Roger Summons, then with the Australian Geological Survey Organization (AGSO), and colleagues found that living cyanobacteria produce a distinct membrane in tiny amounts. There's nothing remarkable about it. It's just a rare molecular signature, called 2α-methylbacteriohopanepolyol. Scientists understand how molecules degrade with time, heat, and pressure. That gave them a specific molecular target to hunt for. If ancient cyanobacteria lived like today's do, they should leave behind a degraded form of this molecule, a five-ring lattice of thirty-one carbon atoms called 2α-methylhopane. So, if the molecule turns up in the rock record, it might indicate the final resting place of the most ancient known cyanobacteria. (The general molecule type hopane, that same five-ring carbon skeleton, may exist buried in rocks in quantities as large as the carbon in all living things.)[33]

Meanwhile, another member of the AGSO team distilled these very biomarkers out of samples.[34] Jochen Brocks, a German chemist who moved to the University of Sydney for a geochemistry Ph.D., and his colleagues, bathed rock samples in solvents three times before they thought any contaminant oil was gone. The appearance of new hydrocarbons persuaded them they were finally looking at uncontaminated rock. The samples were ground into smaller pieces to get at more pristine surfaces.

Every gram of rock yielded just 25 millionths of a gram of research-worthy oils. The 2α-methylhopanes showed up in the distilled oils, which suggests cyanobacteria lived 2.7 billion years ago in what today is northwestern Australia.

The team also published results that offer corroborating, if indirect, evidence that cyanobacteria lived there. If they were there, they would have pulsed oxygen into the air, molecules that oxygen-breathing eukaryotic cells could have cooked into signature molecules—if they existed at that time. So when Brocks found in the samples steroid molecules—essentially ossified cholesterol—he and his colleagues inferred that eukaryotes had evolved by 2.7 billion years ago, a full billion years earlier than the previous estimate. Woody Allen once said, "I don't want to achieve immortality through my work. I want to achieve it by not dying." Short of that, the four-ring network and side chain of carbon atoms in cholesterol might last longer than anything else we have to offer.

Many prominent scientists accepted the evidence suggesting that the biomarkers are 2.7 billion years old. But that's no guarantee. The rocks might still turn out to be contaminated. They might not be. It is impossible to prove the age of a molecule. The scientific enterprise often demands simultaneously the excitement of discovery and relentless skepticism or ambiguity. Jochen Brocks explains that geochemists date old rocks by assembling circumstantial evidence for and against the notion that molecules are not contaminants, that they are native to that layer of sediment. For the biomarkers in the Archean rocks, there is a lot of evidence that they are truly ancient. They've been cooked thoroughly and transformed as geochemists might expect. The scientists found no "modern" (less than 550-million-year-old) contaminants.[35] The chemistry his lab performed is not easily replicated. Brocks says scientists, particularly senior scientists he admires, who can neither redo nor disprove his work, nor accommodate his research into their own hypotheses should just cite the uncertainty of the Archean eon, and make their case for a later rise of cyanobacteria.

Work continues corroborating Brocks's cyanobacterial molecular fossils at other sites. With his students, Roger Summons, now at MIT, has found 2.6-billion-year-old 2α-methylhopanes in southern Africa, which appear even less likely than the Australian biomarkers to be contaminated—a possibility that Brocks himself has been unable to dismiss. The two scientists disagree on the likelihood of contamination in the 1999 samples. In science, that constitutes a reason to work together more closely rather than less so. "Scientists embrace these kinds of disagreements because it forces us to be self-critical and find better evidence; it's the way forward," says Summons. "Ultimately, some bright student will find a different handle to turn and then we will know who was/is correct. Jochen and I don't presently see eye-to-eye on the Archean biomarkers but we specifically asked for him to review our new Griqualand paper because we know he would be our harshest critic."[36]

Two drill cores from the Agouron Griqualand Drilling Project in South Africa were extracted with special care to minimize the potential for contamination. The drills did not require fluids other than water to permeate the lower strata, and the material was examined as quickly as possible after extraction. The two drill cores, about twenty miles apart, contain similar patterns of evidence of cyanobacterial biomarkers, and signs that these hydrocarbons are indigenous to their host rock. In other words, the cyanos probably lived and died there, their remains buried in sediment in the late Archean. Corroborating evidence for steranes appear in these cores as well.[37]

If cyanobacteria lived in Shark Bay 2.7 billion years ago, and molecular oxygen reached a critical level in the atmosphere between 2.45 billion and 2.2 billion years ago, 250 million to 500 million years must be accounted for.

"Oxygen oases" may have simmered over 250 million years, according to one hypothesis, based on changes in the carbon-12 to carbon-13 ratio in western Australian sediment over that period. The oxygen might gradually have collected and reached some threshold 2.45 billion years ago, when rocks show their earliest traces of systemic change.[38] Or, the oxygenation of the atmosphere might have occurred as a flip, rather than a slow leak.[39] Computer modeling guided by physical evidence suggests that the gas concentration in the atmosphere was steady, until the accumulation passed a threshold beyond which ozone (O_3) allowed molecular oxygen (O_2) to accumulate closer to the ground with speed. A steady output of scientific papers every year shapes the current picture of when and how quickly the oxygen came.

If the 2.7-billion-year date for biomarkers turns out not to hold, these studies are addressing a problem that doesn't exist. Cyanobacteria might have evolved later, leaving no 250-million to 400-million-year gap that needs filling. Joseph Kirschvink of Caltech doubts the age of Brocks's biomarkers and argues the cyanobacteria might only have needed a million years to oxygenate the atmosphere.[40]

Carbon in living things and the oceans couldn't have absorbed enough oxygen to keep it from accumulating in the air. Something must have been absorbing oxygen if it took 300 million years from the advent of cyanobacteria to evidence of atmospheric oxygen. The Earth offered many other sinks for the pollutant to hide in. Sulfur minerals show change beginning 2.45 billion years ago. Just as life prefers the light carbon isotope carbon-12 over carbon-13, oxygen prefers to react with certain sulfur isotopes. The Sun's ultraviolet rays, high in the atmosphere, decomposed sulfur compounds independently of four main sulfur isotopes from the beginning of Earth's history. The material precipitated to the ground reflecting their natural isotope abundance, collecting into compounds, such as pyrite, or fool's gold. But 2.45 billion years ago, this mass-independent sulfur faded away. The lightest sulfur isotope reacted with oxygen to make sulfate aerosols—acid rain—instead of dropping to Earth in the form in which they erupt from volcanoes. By 2.32 billion years ago, the mass-independent fractionation signal in pyrite was gone, making this date the conventional threshold for oxygenation of the atmosphere.[41] The disappearance of certain iron and uranium minerals also supports the ascendance of oxygen around 2.4 billion to 2.2 billion years ago.

Once sulfur and iron absorbed their share of oxygen, it accumulated in the atmosphere of methane, CO_2, and other gases. David Catling and Mark Claire succinctly stated what happened next. "Atmospheric O_2 and CH_4 annihilate each other."[42] The combustion wasn't dramatic like a blue stove-top flame is, but the results were the same. The two gases reacted into carbon dioxide and water. The loss of methane might have occurred over a million years, around 2.32 billion years ago.[43] A methane-to-carbon dioxide greenhouse switch would have caused a rush of energy out of the Earth system that dropped the planet's temperature 10°C.

Next the CO_2 washed from the sky, weakening the greenhouse further. The destruction of the methane greenhouse was the first shoe to drop in a climatic double whammy. It expedited the long-term removal of carbon dioxide from the atmosphere and seas for burial in sediment. The raindrops beat against mountains and exposed rock. Raindrops absorb carbon dioxide on their way down, making them slightly acidic. The acidic rain freed calcium and magnesium ions from silicate rocks and washed them away, down mountain streams, to rivers. The once-atmospheric carbon and once-geological calcium and magnesium poured into the oceans. Their concentrations built far beyond the point of saturation. Once the water holds twenty or twenty-five times more calcium than it can absorb, the concentration overwhelms the energy barrier preventing ions from reacting with each other.[44] Calcium and carbonate formed calcium carbonate (limestone) and fell to the seafloor, removed from the oceanic-atmospheric carbon cycle for a life "whose monotony cannot be thought of without horror," as Primo Levi put it.[45]

The main driver of the carbon cycle and climate through geological time is plate tectonics, the continuous shifting of the Earth's landmasses into new surface patterns of continents and oceans. As continents move, or new ones surface, they offer fresh targets for this silicate rock weathering. This process is most responsible for clearing carbon from the atmosphere over millions of years. This process can drive the temperature down, as carbon migrates to sediment, and subduction takes it back to the mantle and eventually back out of a volcano.[46] David Des Marais of NASA has estimated that underground carbon burial increased episodically, in roughly 100-million-year spurts from 2.5 billion years ago, to just before the end of the Precambrian epoch, less than 600 million years ago.[47]

That's how the other shoe dropped. Rock weathering washed carbon dioxide out of the air and it sloshed to the sea with minerals freed from the continents. Without its methane-CO_2 greenhouse, and a Sun too weak to meet the Earth's energy needs, the planet froze over in a series of large glaciations, punctuated by a catastrophic "snowball Earth" event. Oceans froze perhaps 300m thick at the equator and stayed that way for many millions of years. Half-mile thick ice sheets sat along the tropics. Many living things were wiped out. "It was a close call," writes Joseph Kirschvink of Caltech.[48]

Despite its glaciers covering the oceans, the cold climate would not have affected the output of volcanoes or deep-ocean vents. They continued to spew materials into the air and seas. Some of this material is nutritious for cyanobacteria, mostly iron and phosphorous. Phosphorous is particularly important. Its scarcity in any ecosystem is the limiting factor on growth. After an enormous amount of time (to us)—perhaps 10 million to 70 million years[49]—volcanic emissions built up to create another greenhouse, this time composed largely of carbon dioxide. Since cyanobacteria have ready access to all the water they need, only iron and phosphorous hold them back. So when the ice receded cyanobacteria would have been able to mop up the phosphorous concealed by the glaciers. Accordingly, their numbers blossomed and they flooded the lower atmosphere with oxygen.[50] Evidence for such a scenario lies in the Makganyene formation in South Africa, an enormous bed of manganese that sits atop an even more enormous bed of iron. These mineral beds are what one might expect from a post-snowball gush of cyanobacterial oxygen. Oxygen reacts more readily with iron than manganese. So the oxygen flood precipitated the iron from the oceans, then the manganese on top of it.

The snowball Earth was a catastrophic sieve for living things, separating hardy species from those unable to survive the cold. Evolution has persisted for as long as 4 billion years, through geochemical catastrophes and meteoric holocausts. Perturbations have never gone over the top. They were bad. Many species died. But they didn't sterilize the planet. Too much carbon in the air turns the Earth into a sweatshop. Too little makes it a snowball. Earth has never reached a hot (Venus) or cold (Mars) point of no return. Whatever doesn't kill life makes it stronger.

4 | Inherent Brutality

Predators, Defenses, and the Ocean Carbon Cycle

Competition is inherently brutal, and it inevitably entails conflict.

—Takahiro Fujimoto

The "boring billion" or the "dullest time in Earth history" passed between the atmospheric tumult of the snowball Earth event and evolution's gradual encroachment on the path of carbon through the oceans. The seminal development in this change was the "explosion" of animal life on continental shelves, beginning 542 million years ago at the beginning of the Cambrian period. In this 54-million-year span, all of the animal body plans in living animals developed, within a diversity that the Earth had never seen before or since. Among many innovations, organisms increased control over the production of biological calcium carbonate—better known as shells—with far-reaching ecological consequences.[1]

Gradually, over 300 million years after the Cambrian explosion, life impinged on the nonbiological cycle of carbon through the oceans. Today, the oceans swarm with shelled algae called coccolithophores, living evidence of how even tiny organisms' collective appetite for carbon can redirect the global carbon cycle. These aesthetically appealing organisms, or their near ancestors, appear in the fossil record beginning about 225 million years ago. They are algae plastered with crystalline carbonate disks that approach infinite symmetry. One of them is too small to see, unless you have a powerful microscope that can discern millionths of a meter.

Yet a coccolithophore bloom contains untold trillions of cells and requires satellite photography to take it all in. Every year blooms cover about 1.4 million sq km of ocean, turning waters from the subtropics to Alaska and Norway an exhilarating white-blue. They are so bright, they reflect three times more light than other ocean water, bouncing solar radiation back out into space. The densest blooms can hold 115 million cells per liter.[2]

Coccolithophores, which means "seed stones," and other shell makers built up Eurasia's mountains of chalk over millions of years. Today, they form a vital part of both the marine food chain and carbon's transport from the ocean surface to the seafloor. At ocean-floor depth, less than half the carbon is organic, or from soft tissue.[3] Coccolithophores might produce half the inorganic carbon, or calcium carbonate that falls to the ocean bottom. But their future—and consequently the future stability of the ocean-atmospheric carbon balance that made civilization possible—is troubled because of man-made global warming.

How these eye-catching creatures came to be, and consequently how coccolithophores gained such influence over the oceanic carbon cycle, takes us through the Cambrian period, an explosion of animals and marine ecosystems brought about by dramatic shifts in the Earth's continents, atmosphere, and living things' evolutionary responses to these changes.

Life grew thick skin more than 500 million years ago. It also grew sharp, lancelike protrusions and teeth. Calcium carbonate exoskeletons cloaked predators and prey alike. Trilobites were sleek critters that dominate the fossil record. They have long infatuated amateur fossil hunters and tenured paleontologists. Spindly legs stick out from a symmetrical body of three parallel regions, hence the name, "tri-lobe-ite." Their eyes were made of calcium carbonate—literally made of stone, limestone.[4] They lived all over the world, slurping in defenseless worms from the ocean floor.

The stakes are never even in such predatory pairings. The prey's life is at stake, while a successful defense only delays the predator's dinner.[5] Shelly armor and weaponry fueled an evolutionary arms race that accelerated life through the Cambrian period. The date 542 million years ago has held since 1994, when an international scientific body agreed on the evidence. Stratigraphy, the dating of rock layers, is an old science that benefits

A Coccolithophore, or shelled alga, just 15 microns wide.

from modern precision. Reading rocks and fossils is not rocket science. Nineteenth-century geologists understood the Earth's stratigraphic layers beginning with the Cambrian. Sedimentary layers older than it so eluded them that they did not add any new periods between 1891 and 2004.[6] Geologists divide the world into discrete timescales based on fossils, radioactive clocks, isotopic ratios, or changes in the Earth's magnetic field. There are two *eons*—the Precambrian and the Phanerozoic, the latter of which is Greek for "abundant life," and begins with the Cambrian. These eons subdivide into eleven eras, and many periods, epochs, and ages.[7]

Geology is a global enterprise, and the world scientific community sets out to corroborate evidence from one side of the world with rocks from the other. The Cambrian takes its name from Cambria, England, where the first fossilized animals were found. The modern definition of the Cambrian period is pegged, however, to a fossil worm near Fortune, Newfoundland. A similar fossil is tied to an aberration in carbon-13 isotopes in Death Valley, a likely indicator of mass extinctions just before the Cambrian. Finally, the same carbon-13 shift is correlated to a precise uranium-isotope dating in Oman that put it at 542 million years ago.[8]

For 3 billion years before the Cambrian, cellular goop colonized tidal basins, the shallow continental shelf, black smokers—undersea volcanic vents—and the high seas. Open oceans largely devoid of shell-making organisms may have persisted for more than 90 percent of history. Calcium carbonate production was ruled by the brute, physical forces of the Earth system. Tectonic shifts pushed mountain ranges out of the sea and yanked others beneath it. The Earth's ever-shifting continental face changed the way winds blew and oceans moved. Carbon dioxide dissolved in water eroded rock, its minerals washing to sea. Temperature plunges caused glacier formation. Sea levels retreated, only to spill into new basins, on new continents, when the ice sheets melted. The ocean carbonate system, the central circuit in the overall ocean carbon cycle, had little or no interference from life.

A practical way to think about the transformation of the oceanic carbon cycle from lifeless to life driven was described by two scientists in 2003. Richard Zeebe and Peter Westbroeck divided marine states into the "Strangelove," "neritic," and "Cretan" oceans. They were building a computer model to explore oceanic calcium carbonate saturation levels over time. This three-part division provides a useful framework for thinking about how life changed the oceans over hundreds of millions of years.[9] Carbonate is an ion, or charged particle, made of three oxygen atoms and one carbon (CO_3^{-2}). It reacts with calcium (Ca^{+2}) to make calcium carbonate, a major repository of carbon in ocean sediment and consequently rock.

With tongues in cheek, geologists reach for Dr. Strangelove to describe past oceans whose life was decimated by extinction events. "Strangelove oceans" reigned for the vast majority of Earth's history.[10] Isotopic studies on rocks as old as 3.8 billion years show that limestone precipitated abiotically in seas when ion abundances grew dozens of times higher than the saturation level. Crystal fans, cement slabs, and other seafloor formations appear throughout the ancient geological record, indicating that calcium and carbonate in super-oversaturated waters bonded and fell to the seafloor. Cubic kilometers of calcium carbonate appear throughout the Earth's crust.[11]

Geologists know little about the continents' shapes before 750 million years ago. They have inferred world maps for periods since that date, based on how iron is frozen in volcanic rock around the globe. Iron mapping suggests that major continental drifting began over 600 million years ago and ushered in an extended era of wild climate and ecological instability.[12] By 580 million years ago, rain washed minerals from stubby continents that many, many ages later would drift into North America, Asia, Siberia, and Europe. Gondwanaland, the supercontinent of its day, contained what today we know as South America, Africa, Australia, Antarctica, and India.

Tectonic upheaval set the stage for the Cambrian, but more subtle events occurred as well. An oxygen gush flooded the atmosphere a few million years before the Cambrian, possibly the "fuse" for biodiversity to ignite. The cyanobacteria revolution had occurred less than 2 billion years before, but only before the Cambrian did O_2 amass in a quantity to power big life. Oxygen allowed life to become larger and more creative in two ways. Organisms harnessed free oxygen into biomolecules more complicated than ever before. And aerobic respiration (oxygen breathing) gave living things access to about ten times more chemical energy.[13]

Several studies have confirmed the date for this rise in oxygen. Geochemists in Denmark found that the mix of iron isotopes in Precambrian deep-sea sediments show that no oxygen permeated deep waters by the end of the last Precambrian ice age, 580 million years ago. Another group reported that changes in carbon and sulfur isotopes indicate a rise in O_2 shortly thereafter. Geologists know from analyses of sediment that the first animals appeared about 575 million years ago. That means 5 million years must have been enough time to give animals breathing room. The Danish research suggested that these productive levels of O_2 remained steady, and aided the rise of bilateral mobile animals 25 million years later. By 550 million years ago the Earth bore critters ranging in size from centimeters to a meter.[14]

The Cambrian may have been the seminal event in the history of animal life, but it wasn't the enabling event. Before we leave the microbial world behind, it is worth remarking that every cell in every human, animal, tree, fungus, or alga has three or four bacterial ancestors. We like to think that we are different, that our big brains, bipedalism, and opposable thumbs separate us from the rest of life, which they do. But our development and our health is wholly dependent on life forms smaller, but no less "high" than us.[15] We are the product of a series of microbial mergers and acquisitions so improbable the origin of life looks easy. Every cell in our bodies and in every flower, fungus, and prairie dog contains bacteria abducted and enslaved long ago.

The capture and integration of one cell by another is called endosymbiosis. Nearly all eukaryotes have little organs ("organelles") called mitochondria. These cellular energy centers descend from purple sulfur bacteria, inhabitants of stromatolites in Shark Bay. This class of bacteria has made its living for as long as 3 billion years by using oxygen to burn carbohydrate fuel. Deep in the evolutionary past, some oxygen-breathing bacteria became engulfed within anaerobic cells, which needed help thriving in an atmosphere of increasing oxygen. These bacteria are the ancestors of our cellular power plants. The evidence is that bacteria and mitochondria share much of the same DNA.[16]

Another merger and acquisition occurred much later, during the boring billion. Phytoplankton—basically algae—evolved chloroplasts, which are direct descendants of cyanobacteria. The green of lawns and forests does not come from homegrown plant photosynthetic technology. The skill comes from cyanobacteria that were abducted more than a billion years ago and never freed. Red algae evolved in parallel, also with cyanobacterial metabolism. Coccolithophores make their carbohydrates from sunlight, but deploy slightly different light antennae than their distant cousins, the green plants. In fact, coccolithophore genomes reveal them to have evolved away even from red algae.[17]

The question of how organisms learned to make calcium carbonate shells turns out to be secondary. First, they learned how *not* to make them. With rivers washing all that calcium, magnesium, and dissolved carbon dioxide into the sea, it must have been difficult for goopy Precambrian life to avoid inadvertently becoming platforms for crystal growth. Cells and tissue became accidental catalysts for calcium carbonate formation.[18] Imagine waking up with limestone growths about your neck and shoulders. You might have two choices. Either slather yourself with Vaseline so no more can grow, or let it ruin your chances to lead a healthy life and reproduce.

Organisms stood a better chance of making it through the sieve of natural selection if they were endowed with some mechanism for inhibiting mineral growth. Survivors may have had an ability to secrete some mucouslike, anticalcification defense.[19] Once secreting organisms established their ability to survive, their mutation passed down the generations.

Descendants inherited this skill. Sliming calcium carbonate growths became standard "off-the-shelf" biochemistry, shared by organisms that diverged from common ancestors tens of millions of years before.[20]

Knowing how to prevent something is a short step away from using it to your advantage. Sometime before the Cambrian, shells became an evolutionary benefit. Algal mutants showed up in calcium carbonate garb. Predators gobbled up the mutants' unarmored relatives, leaving the best-adapted organism to reproduce. Shells across the carbonate shell-making species share few visible traits. But many had in common biological pathways that originally prevented calcification. Only later did they bring the calcium and carbonate ions together for their benefit. Eukaryotic organisms developed carbonate shell making independently at least twenty-eight times.[21]

The earliest calcium carbonate fossils date to about 550 million years ago. Shrink a stack of waffle ice-cream cones to a length of 3mm and you'll have an approximate image of the humblest carbonate fossil, *Cloudina*. Their brittle shells housed wormlike creatures. They probably fed off the nutrients provided by either stromatolites or algae at primitive reefs.[22] What the tube dweller within looked like, or whom they are related to is unknown. Geologists have discovered hundreds of these nesting conical skeletons since the early 1970s. They've turned up from Namibia to Death Valley. In China, *Cloudina* tells geologists its most important story.

Cloudina is notable among paleontologists primarily for not being eaten. They eluded predators by retreating into their tubes, the way an ungloved hand disappears into the sleeve of a winter coat. Fossilized shells—rather, holes bored into them—indicate that something considered *Cloudina* a tasty treat, worth rattling out of its armor sleeve. Of one hundred fossilized tubes discovered in China, more than twenty bear the marks of uninvited guests. Whatever predator bored 15–85-micrometer holes into these shells left no other trace behind.[23] Predation is an action, fleeting, and therefore much more difficult to capture in rock than on a 16mm camera or digital handheld. That's why *Cloudina* is important. It's the oldest record of this deadly version of hide-and-seek. Corals and shells eventually transplanted abiotic cement sea fans and stromatolites. As carbonate skills spread, and the organisms within them thrived, evolution's appetite for carbon transformed the ocean's overall carbon cycle.

Carbon-13 isotope patterns left in sediment just below Cambrian fossils indicate catastrophic climatic swings. Few if any Precambrian animals survived into the Cambrian period.

The ancient Greek goddess of love, Aphrodite, makes a lovely metaphor for the emergence of life. Her form took shape from sea foam, and she lived among the waves until Zeus called her up to reside on Mount Olympus. Woe is her lover at that time, Nerites, a minor sea god who refused to leave the waters and accompany Aphrodite. As punishment the gods turned him into a mussel.

A "neritic ocean," named after Nerites, hosts shallow, offshore ecosystems, perhaps 200 meters deep, where mussels, clams, and oysters laze, and shrimp and lobster scamper about. For about 300 million years after the Cambrian explosion, life, big life, ventured out on this shelf. In doing so they interrupted the nonbiological process of calcium carbonate formation in the Strangelove ocean. The neritic ocean is a time of transition. Life hadn't yet stabilized the ocean carbonate cycle, and the chemical forces that precipitated limestone for eons remained the most stable pathway. Species entered and left the scene, none providing enough ecosystem stability to become a dependable transporter of carbon to the deep. But life had stepped up its marine invasion, evidenced by fossils literally up and down the geological record.

Geologists recount with restrained passion their trips to Siberia, Australia, or, naturally, East Cambria, England. In *Life on a Young Planet*, Harvard's Andrew Knoll writes about a cliff on the Kotuikan River in north Siberia. This wall, more than 120 meters high, is a document of 15 million years of evolutionary change. Carbonate rocks along the riverbed show an age of about 544 million years. Millimeter-scale, conical calcite fossils dot these lower strata. Life grew larger and more diverse from there, leaving fossils of increasing complexity up the cliff face. Animals enter the record and leave their "tracks, trails and burrows" behind. Higher still, organisms from more than eighty different groups lie in state, their tripartite symmetry marking them as one-time evolutionary success stories.[24]

There are many proposed explanations of the Cambrian explosion. Stefan Bengtson, a Swedish paleontologist, has pointed out that the most widely accepted cause of the Cambrian is "whatever object or phenome-non is under study."[25] John Hayes, a senior scientist at the Woods Hole Oceanographic Institution, broke into light laughter when I asked if any-one was not "completely bamboozled by the Cambrian" and biomineral-ization's role within it, as he had put it. "If they claim not to be, I think we should be very careful with them," he said.[26] Scholarship has simmered in recent years as the scientific community has assembled the constellation of possible factors into some semblance of chronological order.

First, the tectonic shake-ups and climate spins that date from 750 mil-lion years ago are deep geophysical causes of the Cambrian. In discussion of life and climate on any geological timescale, it is difficult to overesti-mate tectonic factors. The plates' motion causes upheaval that reconfig-ures winds, waters, and weather—and consequently, the climate. The single exception, other than meteorite impacts, is man-made global warm-ing, where human industrial activity is changing the climate orders of magnitude faster than plate tectonics and rock weathering.

Second, oxygen flooded the atmosphere within five years after the end of the last Precambrian ice age. A possible reason for this may be the evo-lution of organic material that microbes found difficult or impossible to break down. Tough polymers weren't easily biodegradable, at least not yet. This led to the storage of carbon in organic materials underground as oxygen took to the skies. "Big life" could never have evolved without a flood of oxygen.

Third, any evolutionary innovation implies mutation in genes and in how the cellular apparatus expresses them into an organism's phenotype, or observable traits. Paleontologists search for commonalities among organisms—living and long extinct—and project shared traits backward, onto hypothesized or fossilized forerunners. The genes that determine symmetrical body forms are thought to have developed tens of millions of years before the Cambrian. The so-called homeotic genes, or *Hox* genes, determine bilateral animal body plans. Today, fruit flies, humans, and worms take their shapes from commands encoded in the *Hox* genes. The same genes that put our heads on our shoulders and our legs in our hips do the same for beetles, bears, and lobsters. (On the gene *Hox 1.1*, the difference between a mouse and a fruit fly is the code for one amino acid. Body plans vary widely, but the instructions that put them in mo-tion, the *Hox* genes, largely do not.)[27]

Tectonics laid the groundwork. Oxygen made big life a biochemical possibility. The genes were already in place. Did the shelly arms race between predators and prey ignite the Cambrian?

Probably not, but they gave it a boost. The arms race was already under way. Abundant raw materials for skeleton making allowed predation and protection to become more intricate. Predation is not limited to armored defenses and weaponry, which only a minority of Cambrian organisms had. Size, speed, and chemical weaponry have long aided hunting and escape. Bengtson argues against the notion of a single "trigger" for the Cambrian. Shells and predation certainly had a role in events, but the Cambrian was far from "a calcareous dress-up party," he writes. The safest explanation was a harmonic convergence of factors.[28] Think of a pile of sand growing taller and wider, grains at a time. At one point, a few extra grains cause a cascade. The Cambrian was an evolutionary cascade, its grains biological crystalline carbonate, oxygen, predation, climate, genetic developments, tectonic shifts, and myriad other considerations.

Predation encouraged the evolutionary expansion and innovation of carbonate shells. High predation correlates with diversification within ecosystems.[29] With so much life scampering about, from microbes to trilobites and beyond, the oceanic food webs became a sort of carbon pinball machine. Some organisms dropped their carbonate remains wherever they lived. Some were eaten. Their carbon passed from animal to animal, either staying in the water column food chain or falling to the floor as sediment.

Shells dropping straight to the seafloor or ending up there after passing through the food chain speak to our main interest—the growing effect of evolution on the carbon cycle. In the neritic ocean, calcium carbonate makers took on a permanent role in the disposition of carbon to the seafloor. But they didn't own the process yet.

A mass extinction separates the Permian and Triassic periods. About 95 percent of species disappeared from the fossil record 251 million years ago. The cause of the event remains disputed. A runaway greenhouse may have caused the die-off. Volcanic eruptions in what today is Siberia spilled lava over an area the size of Europe, to a depth ranging from

400 meters to 3,000 meters. The greenhouse-gas release associated with the eruptions might have ignited warming, and consequently melted the vast, frozen methane beds, called clathrates, in the ocean. If the methane bubbled up, and added fuel to the greenhouse boiler, life would have cooked to near-total extinction, which is what happened.[30] The event occurred in about 1 million years, and life needed 100 million years to replenish its diversity and numbers. Coccolithophores colonized the open ocean for the first time only after the Permian collapse.

Carbon dioxide reduced the alkalinity (or increased the acidity) of the oceans. Dissolved carbon dioxide makes water slightly acidic. Enormous amounts of dissolved CO_2 changes the pH of seawater and thus the conditions for life. Particularly hard hit were organisms that depended on carbonate skeletons. About 88 percent of genera in this group disappeared after the Permian, but only 10 percent of noncarbonate wearing species died. Some organisms able to survive and make their shells in less alkaline waters died when global warming raised temperatures beyond habitability.[31]

The Permian extinction cleared the way for the development of the Cretan ocean. *Creta* is the Latin word for chalk and lends its name to the Cretaceous period, the dinosaurs' heyday and the era of oil production. Most geological periods are named after the site of important fossil finds. The Carboniferous and the Cretaceous are the only two named after material, emphasizing the massive leakage of carbon from the short-term carbon cycle and its burial in sediment.

Through this time, the supercontinent Pangaea began its tectonic breakup, gliding over the course of more than 200 million years into today's familiar position of the continents. By 225 million years ago, coccolithophores inhabiting the open sea surface passed on their shells to the seafloor. These "bugs," as specialists affectionately call them, gave oceans some stability that had been missing during the nearly game-ending climate swings of the Precambrian. With two other kinds of phytoplankton, dinoflagellates, and diatoms, coccolithophores formed a tripod at the base of the marine food chain.[32]

Coccolithophores left their mark in the world's chalk formations, which accumulated from about 100 million to 65 million years ago, growing a millimeter a century. The amount of chalk in the United Kingdom alone is astounding. The White Cliffs of Dover rose as chalk-making organisms died and fell to the seafloor. The chalk stretches beneath the English Channel to France, under Paris, up through Denmark and Central Europe, east to the Crimea and Syria, and south to North Africa.[33] The most abundant of the two hundred or so coccolithophore species is *Emiliania huxleyi*, a relative newcomer, first laying down remains 270,000 years ago, and reaching full bloom just 70,000 years ago.[34] It is named for T. H. Huxley, "Darwin's bulldog," who held forth on chalk as a way to explain Earth history in the 1860s.

Coccolithophores grow their shells on the nutritional equivalent of a shoestring budget. Single-celled eukaryotic blobs don't have a lot of energy or material to make fancy armor. Coccolithophores economize. They are jelly-like unicellular eukaryotes protected by a sphere of overlapping shields, each of which—if not falling intact into nascent rocks on the seafloor—lasts about as long as a week. How they make the shields is an evolutionary lesson in how to achieve the very most using the least materials. Armor is expensive, requiring enormous energy and materials to build and then lug around for the rest of life. Whether the payoff is worth the investment depends on how badly they need it. Coccolithophores apparently need it.

Normally proteins are life's construction crews. But coccolithophores can't afford proteins. They bloom in nitrogen- and phosphorous-poor waters late in the summer. They scrounge for these atoms to use in DNA, RNA, and life-critical proteins. The coccolithophores' proteins don't build the shells themselves, but direct construction through complex carbohydrates called polysaccharides. These molecules, made of hundreds or thousands of atoms, contain only carbon, hydrogen, and oxygen. They pull together the calcium and carbonate ions into crystals. Coccolith making occurs in vesicles where ions bond into calcium carbonate crystals, which are shaped into a ring at what will be the shell's center. The shields take shape as the crystals accumulate, each extending its particular lattice orientation radially, out into a shield. The crystal has its own idea of architecture. The geometry of calcite takes over.[35] As the disks grow, the vesicle bubbles closer to the cell's outer membrane. Once completed, the shell slips out of the tissue, into a defensive position among the sphere's two dozen or so other armored plates, as illustrated on page 70.

Since their first appearance in the mid-Mesozoic era, coccolithophores may have created an important buffer that continuously pumped carbon to the seafloor and stabilized the climate against large hot or cold swings. Their shells today carry about half of the open ocean's inorganic carbon, and more than a quarter of the total carbon, to lower strata.

The bottom of the food web is a crowded place. Some five thousand kinds of flakes and globules bob about the ocean's highest 30 meters. Just a few kinds of organisms drive the whole system and a few species within them. These phytoplankton turn the oceanic carbon cycle, disproportionately to their size, making up less than 1 percent of the Earth's photosynthetic matter but 45 percent of the planet's fixation of carbon into cells. As shell-forming, photosynthetic "bugs," coccolithophores influence the ocean carbon cycle in two important ways. The first, photosynthesis, is old news. They ingest nutrients and cook them in sunlight into higher-energy organic materials. They draw dissolved carbon into their cells, and by extension into the marine food chain, making them a major "sink" for atmospheric carbon dioxide in the ocean. Coccolithophores ingest three to ten times more carbon for use in cell building than they need for shell building.[36]

Coccolithophores are gobbled up by zooplankton, microscopic animals such as pteropods, tiny cousins of clams, and copepods, tiny cousins of crabs. Copepods draw attention from some marine biologists, both for their ecological importance and because they look like science fiction monsters, with their plated body armor, ten legs, and creepy hairlike antennae. Copepods and pteropods digest the coccolithophores' squishy innards. The coccoliths, or shells, have about as much nutritional content as a conch shell. So the zooplankton pack the coccolith plates, as many as 100,000 at a time, into fecal pellets and drop them whenever nature moves them.[37] The material either remains in the food chain, or with the coccoliths providing ballast the pellets fall to the seafloor, removing carbon from the biosphere. Gazillions of organisms packing away poop, and dropping it into the deep ocean, has had a direct and regulatory impact on the climate.

Coccolithophores help draw down the CO_2 human industry pumps into the air, but there's a confusing twist. Shell making produces CO_2. The massive blooms of the most abundant coccolithophore species, *Emiliania huxleyi* may sometimes add more CO_2 to the local atmosphere than they draw down. In the long term, these bugs pack away carbon into the ocean and sediment, even if they make the atmospheric carbon problem worse in the short term. Over long timescales, the net effect is a massive removal of carbon from the atmosphere-ocean system, but it is a net effect.

Seawater chemistry is a tricky thing. The properties of acids and bases are extremely important, as they are throughout biochemistry. Acids and bases are measured on the pH scale. Distilled water and blood have a neutral pH of 7. Acids, which are solutions high in positively charged ions, have pH below 7. Bases, or alkaline solutions, have fewer counts of positively charged protons and pH readings above 7. The oceans are alkaline, with a pH of about 8.1. Man-made global warming is making them less so. The ocean's pH has dropped about 0.1 pH units since the Industrial Revolution. That's about a 30 percent change in ocean chemistry. The pH scale is logarithmic, like the Richter scale for measuring earthquakes.[38]

The absorption of man-made CO_2 emissions into the oceans makes them less alkaline. Ken Caldeira of the Carnegie Institution for Science coined the phrase "ocean acidification" to describe the effect of a stronger CO_2 greenhouse on the Earth's waters. Since shell formation is dependent on pH, changes could threaten the viability of shell-making species that keep the ocean's carbon cycle and consequently the entire climate stable. Ocean acidification is one of the biggest threats of CO_2 emissions.

Scientists have more work to understand how coccolithophores work today, before they can better predict how global warming will change their viability. It's possible that some species are more sensitive to change than others. Studies in 2000 showed the effect of lower ocean pH on *Emiliania huxleyi*. The beautiful shells, with their 50-nm-wide spokes, looked decayed and sickly. Other species, much less populous, might be more resistant to change.

Past climate shocks have disrupted the good work of coccolithophores. After the meteor strike that uprooted the dinosaurs' ecosystems, the oceans may have become more acidic and shell-forming microorganisms died off. (Analyses conducted in the 1980s on this topic remain underdeveloped.) After a few years, weather would have washed the debris from the air, down the planetary sink, to ocean sediment. It's thought that the coccolithophores bounced back after the system shock wore off, even if it took several million years. The same is not likely to be the case if ocean pH drops this century as much as scientists fear—up to 0.7 pH units, enough to create a new ocean system to match the planet's new atmosphere.

5 | The Witness
CO₂ and a Tree of Life

Whisper a wish to the bark of the tree.
—Yoko Ono

Ginkgo biloba embraces the sacred and the mundane like little else in our experience. Its most ancient recognizable relatives are embedded in fossils more than 270 million years old. *Ginkgo's* survival into modernity was not merely a matter of luck. It needed an arsenal of molecular and systemic weapons against disease, predators, and time. *Ginkgo biloba* has witnessed most of tree history, the proliferation of land plants, and their transformative effect on the carbon cycle. By any reasonable expectation, it should have vanished into silt long ago. Instead it lives on, playing both Homer and Cassandra, telling great tales of the past and warning of the future. Human beings are the greatest thing to happen to *Ginkgo* in about 90 million years.

Ginkgo trees can be found in Boston, south to Charleston, South Carolina, along the banks of the Detroit River, and west to the Pacific. Nineteenth-century horticulturalists planted *Ginkgo* in cities up and down the northeastern United States, because the trees tolerated coal soot so well. Today, urban planners put *Ginkgo* to work where automotive pollution, storms, and seasonal temperature swings expose human residents to some risk. Trees must be robust to survive Chicago. The Moscow of the Midwest shuts down virtually every year after a blizzard. Officials cordon off neighborhoods where wind knocked trees into power lines. In institutional memory, the city's senior forester told me, no resident has ever called to say the storm toppled a *Ginkgo*.[1]

One *Ginkgo* withstood the most violent weapon ever deployed. The U.S. Army unleashed the atomic bomb over Hiroshima shortly after eight A.M. on August 6, 1945. More than eighty thousand people died instantaneously. Many thousands more died of radiation sickness within weeks, months, or years. Bamboo trees five miles from the epicenter ignited and burned. The explosion cleared all brush and herbs within a kilometer. No living thing grew back within 700 meters of the explosion.

Yards beyond that limit, the Hosenbou Temple lay utterly in ruin. In the spring of 1946, something remarkable and unexpected happened. A tree in front of the temple sprouted from its roots, despite the destruction of its trunk.[2] Long a cherished Japanese cultural and religious symbol, *Ginkgo biloba* one morning became the ultimate symbol of renewal.

The mystery and defiance embedded in *Ginkgo*'s history reached its symbolic apogee that spring morning in 1946. It's a tree that has taken more than 270 million years to prune—a longer history than any other living tree species. *Ginkgo* has many stories to tell. Beyond its historical and cultural significance, particularly after Hiroshima, *Ginkgo* is more than a living fossil. It is a living symbol of survival through extremes of conditions and time.

For generations biologists have used trees as visual symbols for evolutionary relationships. In the century and a half since *Origin of Species* came out, biologists have redrawn its one diagram, a schematic tree of life, countless times for countless branches of life. Darwin did not invent the tree of life diagram, but he articulated it better than anybody else. "As buds give rise by growth to fresh buds, and these, if vigorous, branch out and overtop on all sides many a feebler branch, so by generation I believe it has been with the great Tree of Life, which fills with its dead and broken branches the crust of the earth and covers the surface with its ever branching and beautiful ramifications."[3]

Darwin's tree of life first appeared in an 1837 notebook, filled with his observations from the previous six years' travel on the ship *Beagle*. After *Origin of Species* was published in 1859, he fought to plant his diagram in the minds of his peers and opponents. Just seven years later, naturalist Ernst Haeckel published his own famous trees of life. Tree builders were off to the races, already reaching for Darwin's ultimate goal: "The time will come, I believe, though I shall not live to see it, when we shall have very fairly true genealogical trees of each great kingdom of nature."[4]

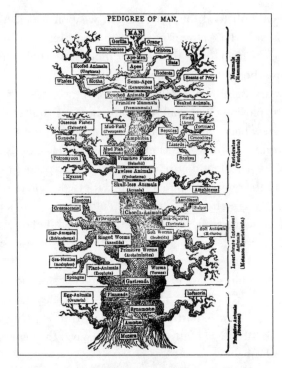

Ernst Haeckel's 1866 illustration "The Pedigree of Man,"
a human-centered early attempt at a "tree of life."

Phylogeny is the search for evolutionary relationships. The common traits that organisms share via a common ancestor are called homologous characters. The more homologous characters two species have in common, the closer the organisms are related. Before molecular biology matured, biologists had only phenotypes—observable traits—to compare. Similar traits don't necessarily imply an evolutionary relationship. Analogous traits appear similar but come from independent lines.

This practice remains central to establishing evolutionary lineages. The mess of data collected—anatomical, metabolic, protein, genetic— are then organized in a matrix of information, a simple table, with species listed in a column and characters running across the top. This matrix is fed into a computer model that evaluates all of the information against criteria that optimize data into a well-pruned tree—wherever possible.[5]

In the 1980s, biologists started to discover specific genes in places that surprised them. Bacteria genes turned up in genomes of Archaea, and vice versa. Far-flung bacteria showed relationships impossible if life were limited to "descent with modification"—the Darwinian notion of reproduction, mutation, and natural selection. These observations could only be explained if microbes gained and lost entire genes or fragments thereof within their lifetimes. Horizontal gene transfer, as this phenomenon is known, threatened the tree of life as a convention of biology. A tree of life makes sense if evolution moves "up" and "out" through time. Genes can jump laterally, and this "genetic crosstalk" confuses classification. A few scientists have jettisoned the convention as theoretically impossible and practically undesirable. But others see in horizontal gene transfer a challenge to renew evolutionary convention with the goal of fulfilling Darwin's vision.[6]

Darwin's vision for a "very fairly true" tree of life comes closest to fruition among biologists who specialize in land plants. Since 1993, hundreds of scientists from more than a dozen countries working together in a group called Deep Green have sorted through morphological, reproductive, and genetic data to assemble a viable tree for the 300,000 or so species of land plants.[7] The data matrix the Deep Green project uses to reconstruct these relationships is based on about ten criteria. From these characters they extrapolated backward through genetic time, to 475 million years or so, when multicellular, sexually reproducing green algae washed ashore and eventually put down roots.

The closest living relative to this progenitor is stonewort, green algae that calcifies in lakes. Plants "climbed the tree," evolving a field of successor worts—such as liverwort or hornwort. Mosses emerged.[8] By 420 million years ago, *Cooksonia* plants sprouted centimeters tall beside rivers and their flood plains. "Moss-like in stature,"[9] *Cooksonia* is thought to be the first vascular plant, which means it grew tissue—plumbing—capable of transporting nutrients and materials internally.

Fossils and computer modeling of plant physiology suggest that leaves became widespread throughout the Carboniferous period. They needed to, because the "meat and potatoes" of their diet—CO_2—became increasingly rare. The atmospheric pressure attributed to CO_2 dropped by 90 percent.[10] As carbon dioxide levels dropped, and oxygen rose, plants and trees benefited from leaves' efficiency at drawing carbon in through pores, called stomata. *Archaeopteris*, the oldest-known woody tree, reached heights of at least 20 feet, by 365 million years ago. It was a progymnosperm, a forerunner of gymnosperms, which include *Ginkgo* and are characterized by having exposed seeds, as opposed to angiosperms, in which the seeds are embedded within fruit. About 350 million years ago, *Lepidodendropsis* rose, its trunk 6 feet in diameter, with its scaly, narrow leaves.[11] These trees slurped voluminous quantities of water from the wetland below. When that local source dried up, or old age took them, the rigidity of the trunks eroded, and they face-planted, sometimes one into another, to the bog below, where over many millions of years, heat and pressure turned them into coal.

Trees became a biological staple for the same reason cyanobacteria conquered ancient bays and shell makers conquered the oceans for millions of years. Trees solved a problem. They tapped resources previously unexploited, with efficiency. As far as plants are concerned, the atmosphere and their ecosystems contain enough carbon that they can set it aside, in wood. More precious than carbon is nitrogen, which figures prominently in proteins, DNA and RNA nucleotide bases, and some small molecules. Rarer than nitrogen is phosphorous. The abundance of phosphorous in an ecosystem often limits the extent of growth. The ratio of carbon, nitrogen, and phosphorus in terrestrial photosynthetic ecosystems is about 830:9:1.[12]

With those kinds of raw materials, it's not surprising that an organism will thrive if it can shuttle carbon out of the way and maximize its nitrogen and phosphorous intake. And that's what trees do. They store carbon in wood, bound into cellulose and lignin, the mostly carbon material that gives wood its stiffness. Phosphorous and nitrogen stay concentrated in leaves, doing their work in nucleic acids, proteins, and the myriad other biomolecules, and then falling to the ground, ready for reabsorption into new leaves or a new organism.[13] Tree trunks also raise leaves above the forest floor, making more sunlight available to them.

Long before *Ginkgo's* ancestors emerged, vascular plants and trees rewired the carbon cycle. The Earth moves most of its carbon around— into the ocean, down to sediment, rock, and back out to the atmosphere— through silicate rock weathering and volcanic degassing. That's how the Strangelove oceans carried carbon from continents to seabed carbonate. Annually, rock weathering is responsible for a small amount of carbon movement. Over tens of millions of years, it adds up.

Trees gave rock weathering a turboboost. Roots produce acids, as does the decay of organic materials. Roots hold the soil in place, which lengthens the amount of time rocks are in contact with these acids. Plants absorb groundwater, some of which escapes through leaves' stomata, or pores. Transpiring water into the air is a temporary measure since it is to come back down again as rain, which further increases the weathering of rock.[14] Roots are complex ecosystems, their slender appendages living symbiotically with microbes, eroding rock into soil, hastening travel to the sea.

Robert Berner of Yale University explains that the global carbon cycle occurs at two speeds, and the long-term cycle gets short shrift. Berner laid out in his 2004 book, *The Phanerozoic Carbon Cycle*, a model of how carbon moved through the Earth's surface and rock over 550 million years. The photosynthesis-respiration pairing makes the short-term cycle of minor consequence over the million-year increments that Berner studies. This means that weathering, burial, and volcanic degassing are the main fluxes of interest and throughout geological time have dictated climate— until today. Or, perhaps another way of putting this is that the long-term carbon cycle still dictates climate conditions, but humans now exert considerable control over it.

The mass of carbon is conserved on Earth. That means the fluxes of carbon among various long-term reservoirs must add up to the same grand total, the carbon freed from the weathering of rocks rich in organic carbon, carbonate rocks, carbon emitted from volcanoes and seafloor vents, and buried in organic and inorganic sediment. The GEOCARB model runs scenarios based on observed and predicted rates of weathering, carbon burial, and degassing, charting the flow of carbon on a timetable that reaches back 550 million years, just before the Cambrian explosion. The success of the model is determined by how closely results match peaks and troughs of CO_2, as determined by geological proxies— molecular fossils from marine plankton and carbon isotopes in ossified soils.[15] Once calibrated to other data, the models become investigative tools. Simulations and geochemical evidence reinforce and propel each other toward new evidence and lines of inquiry.

GEOCARB shows that something extraordinary happened to supplement the copious burial of carbon by rock weathering, something peculiar to the Carboniferous that has never happened on a similar scale since. Burial in swamps made all that dead organic matter inaccessible to oxygen-breathing bacteria. They would otherwise have munched the cellulose into carbon dioxide and perhaps some lignin, too. The trees shriveled and hardened with their carbon intact. The pressure of more and more earth falling on top of them, and heat radiating from the Earth's interior, cooked these ancient forests into coal. Coal, the type extracted from sediment of, say, the Pennsylvanian series, is the result of organic matter crunched anywhere from one seventh to one twentieth its original height.[16] For the Earth system this meant the more carbon disappeared into the muck the more oxygen was left undisturbed to roam the skies.

Oil comes from a period much later than the Carboniferous and its successor, the Permian. Since the Cambrian, carbon accumulated in shallow, sequestered bays off Gondwanaland, the nearly hemisphere-size mash-up of now-fragmented landmasses. One result is the modern Middle East, which sits atop a geological harmonic convergence without peer and holds more than 60 percent of the world's remaining petroleum. When trilobites scampered about the shallow post-Cambrian seas, the Arabian plate lay at the northeastern edge of Gondwanaland. Melting glaciers flooded recessed lowlands with water—today's central Saudi Arabia. Marine life precipitated with abundance, leaving sediment that would form rich, 75-meter-thick source rock. Oil seeped from source rock into geological "traps" as the Earth's internal heat cooked it over tens of millions of years.[17] Interestingly, the Earth has no standard way of producing oil. Each of six oil-forming regions around the world developed on their own at different times, under different conditions. Carbon-rich source rocks were laid down through the Jurassic and Cretaceous periods, but it took a long time to squeeze the carbon out. About 70 percent of recoverable oil dates from the Coniacian age (89.3 million to 85.8 million years ago) and 50 percent since the Oligocene epoch (33.9 million to 23.0 million years ago).[18]

The world's largest oil field, Ghawar, in Saudi Arabia, is younger than its cousins to the southeast. Its reservoir rock is sandstone laid down during the Permian and Carboniferous periods. Over the Cretaceous, down through the Cenozoic, heat and pressure cooked oil from adjoining oil-rich shales into the modern oil field that covers 2,050 square miles—an area larger than Rhode Island stretched out in a north-south orientation. "It's basic geology. You need five conditions to form a large oil accumulation and these things came together in a beautiful manner over a very large area," said Abdulkader Afifi, an Aramco consultant visiting the United States several years ago.[19] The factors range from the topography upon which the source rocks formed to the evolution of organisms whose carbon settled there. An American petroleum geologist put the matter less dryly, "Fortune indeed has smiled on the Middle East!"[20]

The massive burial of carbon earned the Carboniferous period its name. Geologists date it from 359 million to 299 million years ago. It was a bizarre time, when carbon burial peaked and accordingly so did atmospheric oxygen. It topped out at more than 30 percent as compared to our nearly 21 percent. This abundance powered animal metabolisms in ways that today seem nightmarish. The Bolsover dragonfly, a giant insect fossil with 20-inch wingspan, is emblematic of Carboniferous fauna.

As woody plants proliferated, the skies changed again. The Sun had aged and therefore put out more heat. Life and geology buried atmospheric carbon the way residents of temperate climates put warmer clothes in the basement when the seasons change. Today, human fossil fuel burning is finishing the job microbial decomposers failed to perform on *Lepidodendropsis* and its peers—mating the carbon and oxygen to produce carbon dioxide and a rapidly advancing greenhouse. Fossil fuel consumption is rebuilding the greenhouse Earth no longer needed—and actually decimated—more than 300 million years ago. Our carbon age burns part of the pre-Carboniferous greenhouse back to the skies, where the Earth— at least our Earth—no longer needs it.

Though *Ginkgo*'s presumed oldest ancestor didn't collapse into soil until after an ice age brought an end to the Carboniferous, fossil leaves reminiscent of it dot the record long before. They are classified in vague categories, which include the optimistic *Ginkgophyllum* and the more realistic *Enigmophyton*.[21]

Ginkgo's is a fairly straightforward evolutionary story, with a twist. It never died. It should have. It does what any good species is supposed to do. It shows up, then leaves fossil snapshots of a growing family, which just as soon trail off into extinction. Hundreds of species qualify as oaks. The top of the maple phylogenetic tree fans out into more than one hundred species. *Ginkgo* has just one living species, the sole survivor of a once-populous family. Most plant species live a few million years. More than 120 million years have passed without *Ginkgo* apparently changing a whit.

Ginkgo is an exception that proves a phylogenic rule-of-thumb. Closely related organisms tend to die out together when their ecosystems change. It makes sense. If nearly everyone living in a town works at the same factory, and the factory moves to another country, everyone loses their jobs. It's the same in biology. If a bunch of related trees all thrive in the same ecosystem, and something happens to it, they are all doomed.

Yet *Ginkgo* has no living cousins. It's really quite strange. The order Ginkgoales left behind fossils of six families, nineteen genera, and many different species. All are extinct except one. *Ginkgo biloba* is the only species in the genus *Ginkgo*, the family Ginkgoaceae, the order Ginkgoales, the class Ginkgoopsida and the division Ginkgophyta, within the kingdom Plantae of the domain Eukaryota.[22]

No one has explained why *Ginkgo* is a survivor, though few have rigorously tried. Does it have some biochemical defense mechanism? Is it all those thick hydrocarbons in the leaves, keeping bad stuff away like a sealant? Geneticists are getting to the plant world, though only parts of the *Ginkgo* have been sequenced at this writing. And sequencing alone reveals nothing about the mysteries embedded in *Ginkgo*'s phenotype. "Is there whiz-bang chemical magic going on? The morphological stasis means something," says Scott Wing, director of the Smithsonian Institution's Paleobotany Division. "I just don't know what it is." He's not alone. One factor undoubtedly is the tree's ability to regenerate by sprouting from its base, as the Hosenbou *Ginkgo* did in the spring of 1946.[23]

The oldest fossil with more than a passing resemblance to *Ginkgo* was dug from the Earth by a Russian geologist in 1914. Geologists found so many *Ginkgo* fossils between the 1880s and 1960s, that "ginkgology" entered a naming crisis. It got so confusing that in 1968 one scholar implored peers to cut it out. "There is a strong tendency among paleobotanists to create new species almost *ad infinitum*," wrote Hans Tralau, who, reviewing a century of *Ginkgo* fossil discoveries, found just one instance that did not result in a new species.[24] Every budding paleobotanist who stubbed his toe over a *Ginkgo*-stamped rock called it a new species. There's a lot of *Ginkgo* out there to stub your toe on.

The fossil leaves are a silvery brown or black. Really good ones are a flaky translucent brown. Sometimes a geologist in the field will crack open a fissure and find a waxy fossil that sunlight has not brushed in 60 million years. Despite eons in hibernation, a sudden gust can make these

This Ginkgo adiantoides *fossil, unearthed in North Dakota,*
dates to the late Paleocene epoch, which ended about 56 million years ago.
The fossil is virtually identical to
modern Ginkgo biloba.

ghost leaves flap about in the wind, as if they suddenly awakened from a nap. *Ginkgo* leaves have a thick, waxy cuticle. This wax—hydrocarbon chains 27–33 carbon atoms long—may protect the lifeless leaf from total degradation. Peter Del Tredici, senior research scientist at Harvard University's Arnold Arboretum and a world expert on *Ginkgo biloba* put it squarely. This is a plant that thrives on abuse.[25]

Ambiguity shrouds the earliest fossils. Originally, eager paleobotanists put *Ginkgo*'s debut at the end of the Carboniferous period. Genetic analysis may yet bear out that estimate, though even if some proto *Ginkgo* emerged, they would not have been populous enough to join the coal-forming marathon. Fossils found indicate the order Ginkgoales "suddenly appears" thereafter, in the Permian period.[26] The oldest certain fossils of the living genus were found in the middle of the last century in southern Soviet Asia and date to the early Jurassic period, perhaps 175 million years ago. The geologists inferred from those leaves an Asian origin of *Ginkgo*. Studies since have obscured *Ginkgo*'s geographical origins. The trees spread outward, into many families and species, but didn't leave any obvious geographical path showing their radiation.[27]

Jurassic parks were filled with *Ginkgo* between 200 million and 146 million years ago, and surely Triassic (251 million to 200 million years ago) and early Cretaceous (146 million to 66 million years ago) ones, too. A 100-million-year gap in *Ginkgo*'s otherwise prodigious fossil making puzzled scholars for many years. Chinese paleontologists found a missing link in 2003, when they reported a 121-million-year-old *Ginkgo* fossil. Reproductive organs reveal it to be virtually identical to *Ginkgo* trees on urban street corners. This fossil suggests the tree has remained largely the same for twice as long as the dinosaurs have been dead.[28]

Eventually climate shifts caught up with *Ginkgo*.[29] Beginning 90 million years ago, their abundance in the fossil record shrinks. That's more than 25 million years before the dinosaurs' meteor hit, a fact that hasn't stopped some ginkgologists from suggesting the two events had something in common. The argument goes that dinosaurs stopped eating and spreading *Ginkgo* seeds. But no one has ever found a dinosaur fossil with *Ginkgo* seeds in its belly or in an ossified pile of dino dung.[30]

By 25 million years ago, *Ginkgo* disappeared from the far north. They exited Europe less than 2 million years ago. Southwestern Japan may have hosted them all the way through, along with wild pockets in China. No one knows if the world's last *Ginkgo* forests, in northeastern China's Tian Mu Mountains, are wild or left by human ancestors thousands of years ago.[31] A Dutch traveler named Engelbert Kaempfer brought *Ginkgo* nuts back from Nagasaki, Japan, at the end of the seventeenth century, along with elaborate sketches and descriptions of the flora he found there. Trees he planted in Utrecht still grow there today.

Sedimentary evidence suggests that *Ginkgo* thrives in "disturbed environments"—riverbeds prone to flooding or land with rapid soil turnover.[32] Usually, plants that thrive in such environs grow quickly, reproduce asexually (which *Ginkgo* can), and never grow big (which *Ginkgo* does). *Ginkgo*'s affection for disturbed environments explains the tree's success in wooing nineteenth- and twentieth-century urban planners. Sidewalks are as disturbed as any environment. *Ginkgo biloba* became the urban sidewalk's perimeter defense against pollution, heat, cold, disease, and insects. *Ginkgo* ascends columnlike from its block of soil in the pavement, so street managers needn't worry about it growing over the sidewalk. The lowest branches of a maturing tree are high enough for the tallest pedestrian to clear, which saves time and money on pruning. It's efficient. *Ginkgo* tolerates the salt scattered on icy roads. Predatory insects apparently have little or no appetite for *Ginkgo*. This natural insecticide led readers in some Eastern cultures, where *Ginkgo* is native, to use its leaves as bookmarks.

It's also a lovely tree, once it reaches maturity. Young *Ginkgo* trees have a touch of the Charlie Brown Christmas tree. Rigid, awkward branches stick straight out from wimpy trunks. Imagine leafy billiard cues plugged haphazardly up and down a bark-wrapped flagpole. *Ginkgo* is the ugly duckling of trees, taking as long as one hundred years to reach glory. "There are few trees whose youth gives so little indication of future splendor," C. S. Sargent, Harvard University's top botanist, wrote in 1897. "So badly does it blend with American surroundings that a great landscape gardener, knowing only young trees, declared that it could have no place in our landscape planting."[33] Within months of writing those words, the Chinese envoy to the United States commemorated the life of Ulysses S. Grant by planting a *Ginkgo* tree behind the eighteenth U.S. president's neoclassical mausoleum on the Upper West Side of Manhattan. Today, as Sargent would have known, that tree rises 80 feet in the air, within a fenced-in area and commemorative plaque. For *Ginkgo* to truly strike awe, trees need to grow for two to four times longer than the United States has existed. Many landmark individuals in the Far East are hundreds of years old, with the oldest known living tree more than 2,000 years old.

Urban foresters have the same trouble with *Ginkgo* that camp counselors have with precocious teens. How do you stop them from having sex? *Ginkgo* reproduction is a regular "problem" for cities, and has been for more than a century. *Ginkgo* nuts grow a fruity exterior that smells more than passingly like dog doo and contains a dermatitis-causing chemical. In Washington, D.C., female tree trunks are spray painted with a round yellow mark, the botanical equivalent of the scarlet letter. Historical newspaper accounts and chats with urban foresters make it seem as if residents learn for the first time every generation that *Ginkgo* nuts—a delicacy if cooked properly—just smell awful.

Ginkgo sex is more familiar than you might expect, assuming that you expect it to be not at all similar. "It's the same story," says William Friedman, of the University of Colorado. "It's about a sperm meeting an egg."[34]

Asexual reproduction carried life through 2.5 billion loveless years. Sexual reproduction provided a less efficient but friendlier way of passing genes through generations. The first land plants, the algae similar to modern stonewort, already reproduced sexually. If you're out walking around on a rainy day, and you see pollen splayed on the sidewalk, you actually are walking through thousands of moss sperm that are swimming around trying to find moss eggs. Be careful.

Ginkgo retains its algal sexual origins—its sperm swim for eggs. Only they do it 70 feet above the ground. Size matters. A large male *Ginkgo* is better positioned to have the wind carry its pollen grains through the air, toward an ovule on a female tree. The ovule in the female tree catches and absorbs the pollen. Inside, it activates the pollen into sperm. *Ginkgo* has some of the largest sperm on planet Earth. At 86 micrometers in diameter they are more than three times larger than human sperm.[35] They swim differently, too. Human sperm propel themselves forward by wiggling and whipping from the stern, a throwback to the days of unicellular bacterial flagella. Not so in *Ginkgo*. Sperm have thousands of flagella, each positioned toward the bow, yanking itself forward through a viscous liquid released in the ovule. By August or September, the seed is fertilized, and embryo-bearing nuts drop to the ground in flesh foul smelling to most urbanites. For the lucky embryos, some animal—understandably mistaking the smell for carrion—eats it, and passes it in its nutrient-rich manure, which readies it to take root the following spring.

As global warming heated up through the 1980s, 1990s, and the first decade of the twenty-first century, paleobotanists turned to *Ginkgo* for insight. *Ginkgo*'s leaves are not only exquisite, but they age well, and the stomata through which *Ginkgo* leaves inhale carbon dioxide are a key clue to our past, present, and possibly future climate.

Scientists have spent the last several years studying the relationship of *Ginkgo* stomata growth to atmospheric CO_2. Sensitive instruments measure the level of CO_2 with great precision. By counting the stomata on leaves grown under CO_2 conditions known with great precision, researchers can develop a sense of the relationship between leaves and their carbon source. In the laboratory, they can also control the CO_2 content, to see if leaves gain or lose stomata when the CO_2 increases or decreases. *Ginkgo* is not the only subject of study. Research and experiments conducted on many plant species show that increased CO_2 leads to either narrower pores or growing fewer of them. Paleobotanists surmise past climate conditions from counting stomata.[36] The relationship appears to be genetically coded, but that research is preliminary.

When *Ginkgo* grows in an atmosphere high in carbon dioxide, it needs fewer stomata, per square millimeter of leaf, to satisfy its appetite. Similarly, when the carbon dioxide content of the atmosphere is low, leaves produce more stomata—more workers to mine a scarcer resource.

Understanding how leaves react to varying CO_2 is the first step. Next, researchers analyze *Ginkgo* fossils millions of years old, counting stomata under microscopes or on photographs of microscopic images. Stomata counting itself makes for a less-than-dramatic afternoon, but it has yielded some intriguing results.[37]

A team of Chinese and British scientists compared *Ginkgo biloba* leaves collected in 1998 with others preserved since 1924, and with four fossils, the remains of extinct *Ginkgo* species.[38] The scientists hoped to find a correlation between the number of stomata in a given leaf and the known CO_2 content of the air. At the turn of the twenty-first century, the atmosphere contained about 380 ppm (parts per million) of carbon dioxide, and is increasing by about 2 ppm a year. The study found that the number of stomata per square millimeter in *Ginkgo* leaves decreased by nearly 30 percent since 1924, from 134 stomata per square millimeter to 97 stomata per square millimeter. Low counts on the Cretaceous and Jurassic fossil leaves corroborate other proxy geological evidence of high CO_2 levels in those periods.

Researchers looked to fossils of four *Ginkgo* species, from the period from 58.5 to 53.4 million years ago, a virtually identical twin to modern *Ginkgo biloba*. They found the stomata must have grown in a greenhouse with CO_2 values of 300–450 ppm, through a major extinction event, the Paleocene-Eocene thermal maximum (PETM). The results are too low for CO_2 to have single-handedly caused the PETM, one of Earth's five great extinctions (not including the current, human-induced sixth extinction). But that's an easy problem to have. Many factors can contribute to a hothouse. A methane burst from the oceans may have triggered the PETM. Much harder to explain would be an icy period with high carbon dioxide content. Fortunately no one has ever observed any situation like it.[39]

Other sets of stomata data reinforce the correlation between temperature and carbon dioxide that occurs throughout the geological record. Anthropogenic global warming is unique in that we know with certainty what is causing the additional carbon contribution to the atmosphere— industrial combustion and deforestation—and therefore its effect on temperature. The correlations scientists observe today through precision sensing are complemented with indirect evidence—proxies—of temperature and carbon dioxide from the past. Oxygen isotope ratios indicate past temperatures, as reflected in certain marine fossils. Stomata indices in fossilized leaves of extinct plants, and *Ginkgo,* suggest the carbon dioxide content of past greenhouses.[40]

The relationship between *Ginkgo* stomata and previously projected CO_2 levels of past climates go haywire once the carbon content of the atmosphere reaches much higher than today's levels. Leaves become less sensitive to CO_2 change above 500 ppm. We're above 380 and rising— about 430 ppm including the effects of the other greenhouse gases. That's one problem, and limits their utility in reading the past. In its fourth assessment report, the Intergovernmental Panel on Climate Change devoted a chapter to paleoclimate evidence of past climates for the first time. The authors signaled the relevance of ongoing studies. But they cautioned that for the time being, looking back farther than a million years, let alone 50 million to 200 million years, offers too many variables to apply lessons to Earth's immediate future. Once time frames stretch out in the millions of years, geological processes overtake biological signals. The authors write, "Changes in CO_2 on these long time scales are thought to be driven by changes in tectonic processes (e.g., volcanic activity source and silicate weathering drawdown)."[41]

However uncertain paleoclimate proxies may be, the science is creative and intriguing, and nonetheless corroborates study of more direct lines of evidence. Shortly after the fourth assessment report came out in early 2007, Robert Berner and Jeffrey Park of Yale and Dana Royer of Wesleyan modeled climate sensitivity—the temperature change from a doubling of CO_2—over 420 million years. To correlate the model runs, Berner matched his results to proxy evidence—biomolecules left by marine plankton and carbon isotopes of ossified soils—assembled by Royer, as a check on the model. They found that the 420-million-year model and the geochemical proxies concur at a 2.8°C temperature rise for a doubling of CO_2.[42] For a carbon doubling to cause less than a 1.5°C temperature rise, extremely unlikely events would have to occur, such as that plants would have enough light, nutrients, and space to absorb all that carbon as it rejoins the atmosphere. An even lower temperature target busted the model. CO_2 moved into negative levels.

Today, *Ginkgo* is on the move again, anticipating a unique episode in its millions of encounters with climate change. Between 1953 and 2000, *Ginkgo*'s growing season changed in Japan, beginning four days earlier in the spring and ending eight days later in the fall. *Ginkgo* knows the world is warming again.[43]

Japan and other nations in East Asia are studded with ancient *Ginkgo*. Many consider it a sacred tree. A 3,000-year-old tree in Folaishan, Ju County, measures more than 12 feet wide at its trunk and 80 feet tall. The Japanese destroyed many trees during their 1941 invasion of China. That and local overcutting for firewood or charcoal had a destructive impact. In 1956, Chinese scientists vowed to protect Tian Mu Shan, the last-known forest with possibly wild *Ginkgo*, though that intention was largely symbolic until the early 1980s. DNA testing has shown that pockets of wild *Ginkgo* still exist. Trees are marked by greater genetic diversity than street trees, which often start out as grafts from living trees—clones—rather than as fertilized seeds.[44]

A simple *Ginkgo* tree stands near the shore of the Detroit River, both a landscape and art installation. Yoko Ono, the avant-garde artist and Beatle widow, recalls her youth in Japan, where the faithful inscribe messages and place them in sacred trees, including *Ginkgo*, near shrines and temples. "From afar, all those bits of papers looked like white flowers," she once said. "The sight impressed me when I was very young, and I've never forgotten." In 2000, Ono donated the *Ginkgo* to Detroit as part of an ongoing "wish tree" project—targeted plantings meant to invoke the ancient practice of leaving one's prayer in a tree. "A *Ginkgo* was chosen even though it is a fairly rare tree this far north," explains Jon Hendricks, Ono's curator. "It is an ancient tree, which has survived for so long under extraordinary climatic change, that perhaps it could serve as a symbol for the need of cities like Detroit to survive."[45]

6 | Body Heat

Running on Carbohydrates and Hydrocarbons

> As a well-known sports announcer would have put it: "ATP is the most underrated molecule in the league today."
>
> —Harold Morowitz

Every third Sunday in May, 65,000 people—world-class runners, committed amateurs, and loons wearing costumes—assemble at the starting line of San Francisco's Bay to Breakers 12-k race. Each year, a couple of hundred competitors strip to their shoes, don yellow cycling caps, and compete as nature intended.

Clothing is a relatively new innovation in history. In the past, human ancestors were as hairy as apes and didn't need clothes. Darwin supposed that tropical heat caused the hair to retreat from skin touched by sunlight, with the curious exception of the head. Human ancestors started dressing up about 50,000 years ago, a date corroborated by archeological evidence and the genetic history of cloth-transmitted body lice.[1] No other organism wears clothes. If they're cold, they just move closer to the equator or closer to the ocean-floor volcanic vent, in the case of some extremophile microbes and other unfamiliar life.

Scientists have long scratched their hairy heads and stroked their beards, wondering about human hairlessness. Dennis Bramble, a biomechanics specialist at the University of Utah, and Daniel Lieberman, a physical anthropologist at Harvard, sifted through previous research, measured and recorded the human body in motion, and studied clues from fossil skeletons.[2] Their conclusion suggests that endurance running should be elevated from a fitness hobby to a critical evolutionary survival skill. And it turns out that hairlessness, which helps humans efficiently emit body heat, is just one of myriad physiological traits—powerful, spring-like leg tendons, muscular rear ends, profuse sweating—that embolden the hypothesis that running played an important role in human development and survival.

At a popular race like the Bay to Breakers, the back of the pack takes twenty minutes or longer to cross the starting line. The crowding restricts the pace to a walk. Hominins parted ways with apes—hominids—and have walked upright for as long as 6 million years. *Orrorin tugenensis* femurs from Kenya suggest to some anthropologists and biomechanics researchers that human ancestors strode on two feet not long after they diverged from the common ancestor with chimpanzees.[3] The *Australopithecus afarensis* skeleton known as Lucy walked on ground beneath modern Hadar, in Ethiopia, 3.18 million years ago. Her legs show evidence for nearly everything you would want in a bipedal body, most notably a sturdy pelvis.[4]

Half of walking is essentially falling. Biomechanics experts call it an inverted pendulum. The walker's center of gravity is lowest in full stride, the back foot poised on the toes, ready to lift. The front heel touches the ground. As the back leg moves forward, the center of gravity lifts. This center is located around the drawstring in running shorts, or sub belly button for the Bay to Breakers nudes. The height of the body's mass-center peaks as the back leg crosses to the fore. The planted leg at this instant extends nearly straight down to the ground, descending from the center of gravity. That's the hard part of walking, lifting that center of gravity. From there, walking is just controlled falling. Gravity lowers the center back down. The walker falls toward Earth, stopped only by the planting of the next foot.[5]

The illustration March of Progress shows a line of sequentially less apelike hominins, each with diminishing hair and straightening posture. The image is so iconic, and is so regularly lampooned, it's difficult to believe it first appeared as recently as 1970, in a Time-Life book called *Early Man*.[6] Inspired by many previous illustrations of human or animal evolution, the picture itself is more disinformation than illustration, closer to a police lineup than a picture of ancestral lineage.[7] Evolution is messier than a march. It is even messier than Darwin's ever-branching tree. At best, the March of Progress confuses chronology with causality. *Homo habilis* predated *Homo erectus*, but did not determine him. Evolution is not a march of progress, nor is it a baton-passing relay race to "now," which is too often confused as an evolutionary finish line. The evolutionary finish line will come when the Sun burns out, if not before.

If Bramble and Lieberman's hypothesis is right, the figures toward the right of the illustration, the later, hairless marchers shouldn't be walking. They should be running. Fossil skeletons provide evidence that running anatomy took shape after 2.6 million years ago. Stone tool artifacts from this period found in the Gona River Valley in Ethiopia suggest that hominins were eating meat, which means they'd figured out a way (i.e., running) to get to felled antelopes, zebras, and pigs ahead of other creatures. This may have helped them adopt the protein and hydrocarbon diets of long-toothed cats and other predators.

Our runner's anatomy emerged around this time. A long neck allowed *Homo erectus* to swing his shoulders in opposite directions of the gait, keeping the head steady, focused on obstacles as he ran toward the "rising column of vultures"—a sign of food scraps left behind. A nuchal ligament connected the back of the head to the base of the neck. Apes don't have this feature. Our windpipes are wider than our nasal passages, allowing mouths to take in oxygen more rapidly. Plus, breathing through the mouth dissipates heat faster than breathing through the nose. Hairlessness provides some natural air conditioning, and our long, slender trunk and thin limbs allow heat to radiate without ever traveling far. An elaborate system of muscles, tendons, and ligaments absorbs the shock of every step, stores some of this energy momentarily, and uncoils, recoiling it into the next stride. Connective tissue, such as the foot's plantar fascia, the Achilles tendon, and the iliotibial band, which stretches from hip to knee, save about half the expected energetic cost of running. Anthropologists have long argued that brain size determined human ascendance, but big butts may have held us over until brains matured. The gluteus maximus is a bulky muscle at the base of the back that helps keep us from toppling over during running. "Why is our gluteus so maximus?" Lieberman asked in a conference presentation.[8]

At the Bay to Breakers starting line, the competitors are eager to run, but most are stuck comfortably, too comfortably, walking along at maybe 1.5 m/s (meters per second). Soon, the crowd thins. The pace picks up. At 2.3 to 2.5 m/s, our entire body switches gears.[9] The pendulum disappears. The torso leans forward. A runner's feet do not touch the ground simultaneously for the next 12km. The competitors do what they—we—were all "born" to do: run.

They're off. The chest cavity expands when the diaphragm muscle contracts. The lungs suck in air to fill the larger space and equalize pressure with the outside. Blood flowing through the lungs from the body carries carbon dioxide—perhaps the same carbon atom that entered the body in spaghetti or energy drinks or whatever else the runner ingested in the previous days. What happens next is emblematic of how a lot of material travels in the body: the path of least resistance. Chemical and pressure gradients push and pull molecules, atoms, and subatomic particles around the body.

Carbon dioxide and oxygen swap places by diffusing through the lung membrane in opposite directions. The concentration of the CO_2 in blood is much lower than the concentration of O_2 in the lung's alveoli. So CO_2 flows from the bloodstream into the alveoli. Four O_2 molecules bond to every heme, a cloverleaf-looking molecule of thirty-four carbon atoms encircling a center of four nitrogen atoms and an iron bull's eye. It rides out of the area attached to the receptor molecule, globin. The hemoglobin structure carries oxygen through the bloodstream to a muscle cell, which needs oxygen and must get rid of its CO_2. So the reverse gas exchange occurs: Oxygen enters the cell and CO_2 turns into an acid and is transported back to the lungs in blood plasma.

Mammals, insects, fish, even green plants, and particularly oxygen-breathing microbes burn carbohydrates and in this way balance the respiration-photosynthesis coupling of the short-term carbon cycle. Combustion is the energy-yielding reaction of oxygen with carbon and hydrogen atoms that produces carbon dioxide and water. Body heat is a waste product of self-sustaining biological combustion, analogous to the heat from a candle or a car engine. Hairlessness helps human cursors efficiently emit the body heat generated by continuous combustion in trillions of cellular power plants, called mitochondria.

Scientists realized that combustion powers mammals at the same time coal combustion steam engines gained popularity. In 1780, Antoine Lavoisier and Pierre Simon de Laplace conducted experiments in which they compared the result of coal burning to that of guinea pig exhalation. The emissions were identical. Lavoisier wrote:

> In general, respiration is nothing but a slow combustion of carbon and hydrogen, which is entirely similar to that which occurs in a lighted lamp or candle, and that, from this point of view, animals that respire are true combustible bodies that burn and consume themselves . . . One may say that this analogy between combustion and respiration has not escaped the notice of the poets, or rather the philosophers of antiquity, and which they had expounded and interpreted. This fire stolen from heaven, this torch of Prometheus, does not only represent an ingenious and poetic idea, it is a faithful picture of the operations of nature, at least for animals that breathe; one may therefore say, with the ancients, that the torch of life lights itself at the moment the infant breathes for the first time, and it does not extinguish itself except at death."[10]

This is an undeniable commonality between nature and machines. Breathing and burning are not disparate parts of our experience, though culture drives them apart. Everyone has a fire in his belly of about 98.6°F (37°C). No combustion is more internal than that. Machines burn hydrocarbons. Animals burn hydrocarbons and carbohydrates. Carbon and hydrogen release energy when their bonds break and they reform molecules with oxygen—whether the carbon and hydrogen start as glucose in a runner's quadriceps, iso-octane in a car engine, or wax in a candle. Sometimes, the lines cross. Pigs have been known to eat coal, converting it to dark, inflammable manure.[11]

Before humans rode machines of iron and steel, muscle drove societies. Humans, horses, cheetahs, antelopes, and camels are powered by mitochondria. All eukaryotic life has, or had and lost, mitochondria. These organelles are power plants where energy stored in carbon-hydrogen fuels are converted into usable form, adenosine triphosphate (ATP). Humans have 300 to 400 mitochondria in every muscle cell.[12] Mitochondria are central to the greatest leap in evolutionary history, from prokaryote to eukaryote. "If the mitochondrial merger had not happened then we would not be here today, nor would any other form of intelligent or genuinely multicellular life," argues Nick Lane, in *Power, Sex, Suicide*.[13]

Glucose burning occurs in two stages, without and then with oxygen. The former, called anaerobic respiration, predated and (obviously) survived the oxygen revolution, given its ubiquity among eukaryotes and microbes. Anaerobic respiration is to oxygen breathing what the Model T is to a Formula One race car.

Anaerobic respiration makes ethyl alcohol (ethanol) as a by-product, the same ethanol molecule that drives both inebriation and "flex-fuel" vehicles. It's the same molecule detected in the dense molecular clouds elsewhere in our Galaxy. In a sense, every human cell is a microdistiller.

Any runner who has experienced "dead legs" knows about lactic acid, which accumulates in muscles when an athlete demands more oxygen than the blood can supply. This limit is called the $\dot{V}O_2$ max. When a runner pushes him- or herself to go faster than oxygen can reach muscles, anaerobic respiration takes up the slack, producing lactic acid. The lactic acid is oxidized in mitochondria only later, during recovery. When elite athletes are tested for their maximum oxygen intake, lactic acid production is a threshold. If lactic acid is forming, the $\dot{V}O_2$ max has been reached and the muscles have switched to anaerobic respiration. It's possible to measure the $\dot{V}O_2$ max with the right laboratory equipment. Research athletes don breathing tubes and hop on the treadmill or bicycle. A computer measures the O_2 intake. To test the lactic acid content of blood, the experimenter takes a small sample from the ear lobe and sticks it into a machine that quickly analyzes the lactase content.

The carbon that started out as glucose enters aerobic respiration as a three-carbon product of glucose broken out during the anaerobic cycle.[14] One carbon peels off to make CO_2. The remaning two-carbon molecule binds to a transport that escorts it into a cycle of reactions nearly universal among aerobes that blasts the rest of the energy from it. It's called a cycle because the same molecules that begin this process of carbon burning are regenerated at the end of it. Most frequently, the cycle is called the Krebs cycle, after the scientist who elucidated it. German biologist Hans Krebs discerned the reaction sequences by studying the path of intermediary molecules through pigeon breasts. Experimentalists found minced pigeon-breasts muscle retained its ability to oxidize fuel after it had been separated from the original beast.[15] The two-carbon remnants flip through the eight reactions of the Krebs cycle, during which they are oxidized into CO_2. The electrons plucked from these intermediaries zip through an electron-transport chain in the mitochondrial membrane, and eventually end up in water, some of which is secreted as sweat. The Krebs cycle (a.k.a. the tricarboxylic acid [TCA] cycle, a.k.a. the citric acid cycle) runs the same reactions—in reverse—as the reductive TCA cycle that Harold Morowitz argues is a virtual fossil of life's earliest metabolism.

The discovery of how mitochondria make ATP, life's universal fuel molecule, is a surpassing example of how observation and evidence can trump all else in science, including professional authority, personality politics, prejudgment, and know-it-all hubris. An outcast from the biochemical establishment discovered the mechanism, called chemiosmosis, by which mitochondria generate ATP. Peter Mitchell lived off personal wealth, at his estate in Glynn, England. In the lean post–World War II years, he didn't think twice about driving a rare Rolls Royce. He wore his hair like Beethoven's, and conducted his research in a wing of the estate.[16] In 1961, he published a controversial paper suggesting that cells build their ATP by damming up the mitochondrial membrane with a concentration of protons. The repelling force of the protons' positive charges against each other channels the energy released during respiration toward a release valve, the protein complex ATP synthase. Just as in ATP synthesis in chloroplasts, ATP synthase bonds phosphate groups onto ADP with these protons, as they come through the chute.

Mitchell's chemiosmosis paper met with howls and outrage among molecular biologists. Wrath, condescension, and cold shoulders subsided quietly over the next seventeen years into stunned recognition that he was correct. Mitchell's 1978 Nobel lecture began with a personal view of his transformation from loser to laureate.

> Of course, I might have been wrong, and in any case, was it not the great Max Planck who remarked that a new scientific idea does not triumph by convincing its opponents, but rather because its opponents eventually die? The fact that what began as the chemiosmotic hypothesis has now been acclaimed as the chemiosmotic theory—at the physiological level, even if not at the biochemical level—has therefore aroused in me emotions of astonishment and delight in full and equal measure, which are all the more heartfelt because those who were formerly my most capable opponents are still in the prime of their scientific lives.[17]

In action, ATP is a carbon-free affair. ATP contains ten carbon atoms, all tied up in rings at the inactive end of the molecule. The energy storage and release occurs when a phosphate, or phosphorus-oxygen group, bonds to or breaks away from another phosphate. Once recharged, it zips off to the muscle proteins actin and myosin that perform work. There, ATP reacts with water, losing its third phosphate group. The energy is transferred to the two proteins, which slide against each other to produce a muscle contraction.

Aerobic respiration isn't particularly efficient. Cells use only about 38 percent of glucose's energy. The rest leaves the body as heat. That's the pattern for aerobic respiration in everything from oxygen-breathing microbes to humans. Physiological studies shed light on which organisms make the most efficient use of this universal biochemistry.

Scientists have investigated how human cursors stand against other species by strapping breathing tubes to horses, antelopes, and pigs. The word "cursor" originates from Roman times, to describe message couriers, though its most familiar meaning today evokes the blinking line that "runs" across computer screens. Mammalian cursors are the animal kingdom's largest runners. Antelopes show enormous aerobic capacity, using about 300ml per kilogram of body weight per minute. Top human marathoners might surpass a fifth that.[18] Pronghorn antelopes are breathers nonpareil. Their wider windpipes, bigger lungs, hearts, and muscles move more material, faster, through their cardiopulmonary system. Studies done from the early 1970s show a direct relationship between the size of a mammal and the volume of oxygen it takes in during peak physical activity. Size tends to dictate the $\dot{V}O_2$ max, yet pronghorns take in three times the expected amount of oxygen for a mammal their size.

Extrapolating from studies on animals big and small, physiologists believe a 77kg mammal—a human—should need about 100ml of oxygen per kilogram body mass, per kilometer. In reality, human runners need more than twice that volume—212ml.[19] That's why it's so surprising that humans are top mammalian endurance runners. The skill must come from something other than efficient oxygen use. In 1984, David Carrier, then a Ph.D. candidate at the University of Michigan, argued in an "incomplete and largely speculative" paper that efficient fuel storage and metabolism—and heat dissipation—might make up for the poor oxygen intake. The paper is intriguing to read nearly twenty-five years later.

Part of the answer, Carrier argued, may be the very thing that slows us down: two legs. Unlike cheetahs, greyhounds, or horses, human breathing is not coupled to the running gait in a 1:1 ratio. Quadruped runners exhale when their front legs hit the ground. The chest cavity physically shrinks, setting up a natural rhythm for exhalation. Most four-legged cursors have optimal running speeds when their gait, oxygen intake, and metabolism fall into harmony. Human runners are much more flexible than cursorial quadrupeds in terms of the how they couple breathing with gait. Humans use a 2:1 coupling stride-to-breath ratio most commonly, but can also access several other ratios, 3:1, sometimes 4:1, depending on their speed and metabolic demand. Humans practically never sustain a 1:1 ratio. We tend to breathe much more slowly when running than quadrupeds of equal body size.

Another oddity of human endurance running is the cost of transport. An elite marathoner uses approximately the same amount of energy per unit of body mass, per unit distance, as when he's jogging. "We still have no good mechanistic explanation for the flat cost of transport curve that characterizes human running," Bramble says. "It is not likely to be about the unusual breathing pattern of humans, although that might be a minor part of the mix."[20]

Two legs slow us down. Famously, the cheetah darts after potential prey faster than 60 mph. Yet if a cheetah sprinted that fast for that long, she would cook herself to death. The jungle cats generate sixty times more heat running than they do at rest. Like most nonhuman mammals, they have far fewer sweat glands per square inch of skin. Heat, not exhaustion, limits their distance. After a half-mile sprint they stop, with or without dinner, because they are too hot to continue. They get rid of heat by stopping and panting. Greyhounds pant, too, both to catch their breath and to expel waste-heat energy generated during running. Stopping is a rather crude way of regulating the heat of biological internal combustion, but cheetahs apparently have few other options. C. Richard Taylor and Victoria Rowntree found in 1973 that cheetahs stop chasing prey after a kilometer or so, because their temperature rose too high, to 40.5°C (105°F).[21]

Camels can cover more distance than horses or humans, and do so under searing desert conditions. That hump (in the case of the one-hump camel) doesn't store water, but hydrocarbon fuel. It's like a runner strapping a box of energy bars to his back. Camels' water-retention prowess is found in the blood. As they use up water, camels' blood thickens. Where human runners would keel over with dehydration, camels manage to push viscous oxygenated blood to cells, where mitochondria await it.

Physiologists focus on three classes of molecules when they calculate food's energy content: hydrocarbons, carbohydrates, and proteins. Hydrocarbons store twice the energy of the other two. This word is often used synonymously with petroleum in the mainstream press, but the body builds and burns them without having to extract them from the ground. A hydrocarbon is a hydrocarbon, whether it's a triple-chain triacylglycerol molecule from a muscle cell or octane from oil. Hydrocarbon describes an extremely large category of chemicals, from the diversity of molecules that fill a barrel of oil to the highly specific fats our bodies store energy in. Biological fats have reactive carbon-oxygen double bonds where the body begins to break them down. A hydrocarbon that has 200 carbon atoms and 402 hydrogen atoms ($C_{200}H_{402}$) has more potential structural varieties than there are estimated to be electrons in the Universe.[22] It's just not a very specific word.

The body burns hydrocarbons at a rate dependent on intensity and duration of exercise. A graph of whole-body fat burning looks like a gentle curve, with fat oxidation peaking between 50 percent and 60 percent of the $\dot{V}O_2$ max and diminishing thereafter as the body relies on carbohydrates.[23] Fatty acids come in three varieties, defined by the quantity of carbon-carbon double bonds along their long chains. Double bonds are easier for the body to chop up and metabolize. Anywhere between four and thirty-six carbon atoms can bind into a fat molecule, most or all of them with single bonds, where each carbon is bonded to two other carbons and two hydrogen.[24] Saturated fats have no double bonds between any carbons. Monounsaturated fats have one carbon-carbon double bond. Polyunsaturated fats contain more than one such bond. Fatty acid chains that have carbon-carbon double bonds consequently have fewer hydrogen bonds. Instead of each carbon bonding to two hydrogen, they make two bonds to each other, and one to hydrogen, leaving them "unsaturated" of hydrogen bonds.

Bonds have important implications for the macroscopic properties of each fat or oil. Vegetable oils are more healthful than lard because the double bonds in oil are easier to break. The difference is visible at room temperature, where vegetable oils are liquid and lard is solid. Long-chain saturated fats, such as those found in a hamburger or bacon, remain solid at room temperature and require a lot of energy to break down. Butter's hydrocarbon chains have a higher percentage of double bonds, which make them softer at room temperature, but still highly saturated. Vegetable oils, such as olive or canola, have an even higher number of double bonds in their triacylglycerol chains, and melt at room temperature.[25] Hydrogenation of vegetable oils turns carbon-carbon double bonds into single ones, and attaches an H to each carbon. When the ingredient list on food packaging claims "partially hydrogenated vegetable oil," it might as well say, "factory-processed synthetic lard."

Protein is an energy source of last resort. Hydrolysis—which means to break apart in water—dissolves the bonds in amino acids, the way a train engineer might unhook cars from each other. The body remakes them into new proteins in ribosomes.

The most user-friendly energy molecule is carbohydrate. Its name gives away its composition. It is a hydrate (H_2O) of carbon (C). Glucose, a carbohydrate, is the body's preferred energy sugar, represented as a six-carbon ring, or $C_6H_{12}O_6$. Many glucose molecules link together, or polymerize, into a starch called glycogen, which muscles and the liver hold in reserve. Carbohydrates yield about half as much energy as fats, but generally require fewer steps to release the food energy and store it in ATP. What's more they enter the body in close to their usable form. Feasting on carbs in the days before a race can double an athlete's glycogen stores and therefore increase endurance.[26] Carbs are also brain food. The brain metabolizes only carbohydrates, and a lot of them. The brain makes up about 2 percent of the body's mass, but consumes 20 percent of its energy.

Energy is the alpha story of all times. Without energy there is nothing. We eat, work, and fight wars to secure energy. The body stores energy in gradients, the pumps that channel protons through ATP synthase, and in high-energy molecules, such as fats, carbohydrates, and proteins.[27] In a physical sense, all matter is convertible to energy. Matter is energy packaged as particles. That's what Albert Einstein expressed in his five-character formula, $e=mc^2$. Energy is mass times the speed of light squared (that's 186,000 mps, squared). Nuclear explosions have no monopoly on $e=mc^2$.[28] Every time you hit the gas pedal or run a mile or burn a candle, an infinitesimal amount of matter is converted to energy. A kilogram of gasoline releases energy during combustion equivalent to just one ten-billionth of its mass.[29]

Energy is commonly defined as "the ability to do work." Longer descriptions reveal how difficult the word is to explain. Light is electromagnetic energy, the spectrum that runs from long, slow radio waves to short, fast high-frequency gamma waves. The middle sliver of the electromagnetic spectrum is visible, or white, light composed of what we see as colors: red, orange, yellow, green, blue, indigo, violet. Kinetic energy is matter in motion. Heat is thermal energy. There's chemical energy in the making and breaking of molecular bonds—that includes all of life. There's the nuclear energy of the Sun and atomic bombs. Gravitational energy keeps you riveted to your seat.[30] To make matters still more confusing, most of these kinds of energy can be converted into one another.

The international scientific community has imposed limited sanity on energy measurement. Sanity is limited because these units bear only indirectly on how we think about cellular energy day to day. Doctors, dieters, athletes, and the Food and Drug Administration measure energy (officially) in Calories, a traditional measure of heat not recognized in the professional standards. As if matters were insufficiently Byzantine, a Calorie, with a capital C, is shorthand for a kilocalorie, or 1,000 lowercase c calories. A calorie is 4.184 J (joules). A Calorie is 4,184 J. Unfortunately, "Calorie" is often misspelled "calorie."

Nonetheless, universal measurements make it possible to calculate the energy in all sorts of different fuels—food, gasoline, bombs. This is a handy skill to have at dinner parties, in case you get in an argument over which stores more energy, 64 fl oz of soda or 2 fl oz of gasoline. (It's close.) The cycling physiologist Allen Lim has calculated that a station wagon driving 240 miles at 60 mph uses the energy equivalent of 1,987 cans of Coca-Cola or 472 Big Macs. Conversely, if you figure that Earth's 6.5 billion people each require an average of 2,000 Calories per day, that's as much fuel as 10, 000 cars would use driving for twenty-four hours at 50 mph.[31]

Carbohydrates, hydrocarbons, and proteins enter the body to be converted not only into fuel but also into structural material and enzymes. Metabolic systems break down food and fix its carbon into all the things that keep an organism up and running. Biochemical pathways also yield the small molecules, life's nimble messengers, which include neurotransmitters, steroids, prostaglandins, and pheromones.

Tyrosine is one of the twenty amino acids that cells use to build proteins. Healthy humans can make tyrosine on their own, which isn't true of the essential amino acids, which we need to eat. We also get plenty of tyrosine in vegetables, legumes, and fish. The body uses it as more than a brick for protein building. It's also the feedstock for some important neurotransmitters. Dopamine lights up the brain's pleasure centers during eating, sex, some illicit drug use, and possibly learning. The adrenal gland refines dopamine into norepinephrine, a neurotransmitter. When norepinephrine loses a carbon atom it becomes epinephrine, a chemical commonly known as adrenaline, a nine-carbon hormone, six atoms of which are tied up in one of nature's ubiquitous aromatic rings.[32]

A competitive runner can't actually "feel adrenaline coursing through his veins," but it is. U.S. scientists began calling adrenaline epinephrine, after a pharmaceutical company trademarked the word "adrenalin." The words are Latin and Greek translations of "hormone on-the-kidney," a product of the *ad* (on) *renal* (the kidney) gland. Scientists mastered the chemical structure of epinephrine more than a century ago, as well as its function. Epinephrine is a particularly well-known small molecule, the first hormone whose structure chemists solved. Epinephrine is the body's alarm, squirted into the bloodstream during "fight-or-flight" situations at the prompting of neuronal signals. The heart rate speeds up, pushing blood through our 100,000-mile-long circulatory network. Blood pressure rises. This makes more hemoglobin available to transport oxygen to muscles. Breathing passages dilate. Epinephrine wakes up the muscles' and liver's stores of glycogen, which is a chain, or polymer, of glucose. Fatty acids are broken down into usable fuel. Epinephrine is one of a team of hormones and miniproteins that alerts cells to shovel glucose to the cells that need it to do work.

If epinephrine is the chemical starter's pistol, chemicals called endocannabinoids represent the finish line. When running became a fad in the 1960s and '70s, sports medicine gurus attributed to epinephrine much more than its status as the body's homeland-security alert. They reasoned that it could be responsible for the "runner's high," which occurs during or after prolonged, strenuous exercise. Within a decade this notion lost favor, as scientists took interest in endorphins, also called opioids for their resemblance to the active ingredient in opium. The "endorphin high" proved difficult to test in a lab.

Scientists in the 1960s found tetrahydrocannabinol, or THC, the active ingredient in marijuana.[33] Two decades later, they discovered its receptors in the brain—all over the brain, the reason they are called cannabinoid receptors. Marijuana's genius is to hitchhike a ride on the human brain's pleasure and pain-deadening centers.

THC stimulates some of the same brain receptors as chocolate. When neither chocolate nor THC are handy, the body makes its own. The runner's high is a product of *endo*cannabinoids, meaning they are made inside the body. Penny Le Couteur and Jay Burreson ask in their book, *Napoleon's Buttons* if the exocannabinoid in chocolate, anandamide, should be made illegal since it produces a neurological response similar to pot.[34] Since the same can be said of endocannabinoids, perhaps running too should be declared a Schedule I controlled substance, or at least taxed and regulated.

For elite athletes the difference between an illegal substance, such as marijuana, and a banned substance, such as synthetic testosterone or epitestosterone, is academic. Both of these chemicals are produced naturally. Both males and females make their own testosterone by knocking a hydrocarbon chain off a cholesterol molecule. All natural human steroids are made from cholesterol. The steranes that turned up in 2.7-billion-year-old Australian molecular fossils have the same ringed carbon frame. *Ster* has the same meaning in the biomarker *ster*anes, chole*ster*ol, *ster*oid, and testo*ster*one. Proge*ster*one regulates the menstrual cycle, and e*ster*rogen and proge*ster*in are key ingredients in birth control pills.

Testosterone, famously, enhances muscle performance, and athletes have long used it for a competitive edge, despite the risk of disqualification and health threat. UCLA's Don Catlin has spent more than twenty years developing tests to spot athletic cheaters. A dead giveaway, his lab has discovered over the last decade, is the carbon isotope ratios. Standard tests for testosterone measure its abundance in urine relative to the closely related chemical epitestosterone, which has no known use in the body. Leading sport regulatory bodies consider athletes with a testosterone-epitestosterone ratio higher than 4:1 to be in violation of doping rules. This is the same test that Floyd Landis failed during the 2006 Tour de France, registering a score of 11:1 that jeopardized his title. Some athletes became wise to the epi-/testosterone test, and began to supplement testosterone doping with epi, to balance out that ratio. Catlin's lab in

2002 published results of a method to sniff out synthetic epitestosterone by detecting deviations in its carbon-13 content. Synthetic epitestosterone, made from starter molecules in yams or soybeans, has fewer carbon-13 atoms among its carbon than the same molecules produced in the body. Animals tolerate the heavier isotope more easily than plants; a low carbon-13 content signals the athlete most likely used performance-enhancing drugs.[35]

In terms of efficiently converting fuel energy into motion, the most efficient form of transportation on the planet is not naked running through city streets. In fact, there's really no analogue in nature for this transport, and it doesn't come from nature. It's not the wheel, but two of them: the bicycle.

Scientists have casually wondered why natural wheels are rare. Bacterial flagella whip in circles to propel the cell forward. Tumbleweeds roll, dispersing their seeds across arid plains. Roundness is built into the scientific name of one tumbleweed species, *Cycloloma atriplicifolium*.[36] Nuts and fruit tend to drop and roll. Wheeliness in that sense helps trees disperse their seeds. Hips and arms roll around in their sockets. Eyeballs roll. The coccoliths of *Calcidiscus quadriperforatus* are round shields, but that roundness has more to do with ease of manufacture than efficient movement. That's about it. The quick answer to why nature has reinvented the eye between forty and sixty-five times, but never the wheel, is simple enough.[37] Eyes are useful everywhere sunlight hits. Wheels are efficient only on hard, flat surfaces. Four legs move through sand or mire with less aesthetic appeal, but greater reliability.

Horse and buggy roads threaded through developed societies by the nineteenth century, but the familiar bicycle—the first revolution in mechanized transport—was introduced in 1885. In that year the Rover safety bicycle, with its standard rear-wheel chain and sprocket drive, was first mass-produced in England. Within fifteen years, 312 factories put out a million bicycles a year. Henry Ford built a steam-powered "quadricycle" in 1896. The Wright brothers made bicycles, as did the Dodges.

The bicycle harnesses the power of mitochondria to make the most efficient known form of transport. Human walking costs about 0.75cal per gram, per kilometer. Horses and camels are more efficient. But on top of a bicycle, a person cuts energy use by 80 percent, and increases speed three or four times.[38] The energy savings come from two features of cycling.

*A member of the Capital Bicycle Club riding down the steps
of the U.S. Capitol in 1884.*

First, riding a bike eliminates the need to spend energy on standing up. Whether walking or running, legs, hips, and waist perform double duty, both propelling the person forward and keeping the spine more or less perpendicular to the ground. On a bike, the seat supports the torso weight, so the rider doesn't have to. Nor is energy expended raising the feet for the next push. Pedals help each other out; when one turns downward, it pushes the other upward. Second, the bicycle directs the powerful thigh muscles into the direction of the rotary.

The abrupt growth in bike technology and popularity may pale in comparison to the mass-psychological effect of biking. A writer for *Horseless Carriage Days* wrote in 1937, "The reason why we did not build mechanical road vehicles before this [1890] in my opinion was because the bicycle had not yet come in numbers and had not directed men's minds to the possibilities of independent long-distance travel on the ordinary highway. We thought the railway was good enough. The bicycle created a new demand, which was beyond the capacity of the railroad to supply. Then it came about that the bicycle could not satisfy the demand it had created. A mechanically propelled vehicle was wanted instead of a foot-propelled one, and we now know that the automobile was the answer."[39]

PART II | The Unnatural

"I'm not sure he's wrong about automobiles," he said. "With all their speed forward they may be a step backward in civilization—that is, in spiritual civilization . . . But automobiles have come, and they bring a greater change in our life than most of us expect. They are here, and almost all outward things are going to be different because of what they bring. They are going to alter war, and they are going to alter peace."

—Eugene, *The Magnificent Ambersons,*
by Booth Tarkington, 1918

7 | Greased Lightning
Carbon and the Car

Thought isn't a form of energy. So how on Earth can it change material processes? That question has still not been answered.

—Vladimir Vernadsky

Why did life take 4 billion years to produce a gasoline-powered automobile but just over a century to fill roads with 800 million of them?

If you compress those 4 billion years into one year, oxygen-producing cyanobacteria set down their earliest traces early in the afternoon of April 30. The planet glazed into a snowball just as the newspaper came the morning of June 2. The Cambrian explosion ignited for a late lunch on November 12. *Ginkgo biloba's* oldest suspected relative fell over on December 7, just before the workday began. Hominin runner-scavengers used stones as cutting tools early evening on December 31. The Neolithic "Great Leap Forward" in Africa occurred six minutes and thirty-four seconds before midnight December 31. Just 1.1 seconds before midnight, there were no cars. When New Year strikes, there are 800 million.

No cars. Boom! Cars. It's not as if you put two cars in a garage you have to keep an eye on them all night.

The geologic blink in which the automobile conquered the world raises an important question. How is it that human ideas, inventions, and innovations emerge in an evolutionary-like pattern but with a speed, nimbleness, and crudeness that distinguish them from biological evolution? The first half of this book unrolled a series of molecules and metabolisms through which the flow of carbon entwined biological evolution with the guiding physical forces of the Earth system. In the second half these topics unfurl in reverse order—combustion, trees, protective armor, pollution, genetics, and space—to emphasize the peculiar relationship between biological evolution and technological progress. They are doppelgangers. Evolution is natural. Technological and economic progress is somehow unnatural. To adapt Nobel laureate Roald Hoffmann's phrase, they are "the same and not the same."[1]

We're a part of the Earth system, a transformative influence at the moment. We emerged from the Earth system, so it's unclear how anything we produce could be unnatural. That said, some working definition of "unnatural" is useful for two reasons. First, we should celebrate the achievements of human organization and intellect. Second, we need to identify the harm we do to the world—and consequently ourselves—so we can try to rectify it. A natural/unnatural distinction is a practical guide for understanding our role in the world, even if the line between them is just a thought experiment conceived to explain the speed of human civilization. Defining the unnatural is ultimately what many economists, technologists, and historians do. They track the patterns by which human knowledge transforms into invention and subsequently acquires economic value. "Despite an enormous amount of research, technology and knowledge have remained a slippery topic for economists," economic historian Joel Mokyr writes.[2] Basic science fuels technological innovation, which propels fossil-fueled human economic development—the novel geological phenomenon that has kicked the global carbon cycle into overdrive.

To understand where technological speed comes from, you have to root out how ideas emerge, how physical objects are manufactured from them, and how this process differs from the molecular machinations of living things.

A working definition of the unnatural should be framed by what "natural" means. For a simple metaphor, the Central Dogma of molecular biology will do for now. Genetic information is coincident with molecular structure, the genome, that twisting scaffold of carbon. There's no separating one from the other. DNA molecules communicate design information with the same nucleotide bases that hold their double helices together, A, T, G, and C. RNA, proteins, and small molecules activate genes and regulatory functions as needed. Life assembles itself and operates because the assembly instructions and raw materials are all interlocking electrochemical parts, all plays of C, H, O, N, P, S, and other atoms. Everything must fit back, through however many intermediaries, to the original step of the electrochemical self-assembly line, DNA. Cells are imprisoned in the twin confines of time and space—their lifetimes and their physical extent.

By contrast, human information is not limited by time or space. The dance parts ways with the dancer. There is no "Central Dogma of human development," but not for a lack of trying. The most well known attempt is a playful thought experiment with the molecular biologists' Central Dogma. In his influential 1976 book, *The Selfish Gene*, Richard Dawkins invents the "meme," a pun on "gene" and "mimetic," to demonstrate that cultural information is replicated imperfectly and undergoes a kind of natural selection, just as actual genes do. As a tool describing the observation that some ideas thrive and some don't, the meme has illustrative utility. But memes elude definition. They encompass everything from ideas to word phrases to tunes—cultural snippets. "Meme" is a successful meme. "Useful knowledge," the late economist Simon Kuznets's phrase, is not. Dawkins's main point with memes is too frequently lost, that human experience, accumulated knowledge, and critical reasoning—all classifiable as memes in one way or another—can temper or conquer the more base, stupid, and brutal elements of our nature.[3] That's the most brightening point of the human enterprise. But it doesn't make memes a Central Dogma for human development. You can't "sequence" a meme the way you can sequence DNA or "sequence"—investigate the historical record of—a patent or a nonfiction book. It's unlikely anyone will produce a more enduring and foundational metaphor of the dynamic flow of knowledge through minds than Dawkins's memes. But memes themselves halt discussion as soon as they start it, because they are so vague and hard to quantify.[4]

A term borrowed from engineering might help explain elusiveness of an explicit parallel between molecular biology and the diffusion of human knowledge. In biology, the DNA-RNA-protein succession indicates a necessary physical synonymy of shape, information, communication, production, and product. "Decoupling" is a good word for the fracture between knowledge and how people manufacture anything from it. It's also a technical word. Engineers call it decoupling when they break down a complicated problem into simpler problems to speed a solution.[5] As a species, we have physically decoupled information—thought, what Darwin called "a secretion of the brain"—and how it is expressed in finished goods, which is to say, by any means necessary—hit it with a rock, snap it in half, build a factory filled with steel stamping robots, sculpt it with your hands, melt it, and mix it with carbon and manganese. Human-made structures take their shape exosomatically, under physical duress, by processes invented on the spot or around conference tables, not biochemical self-assembly lines developed over deep time. Manufacturers take a raw material and "heat, beat and treat" it, under conditions and with chemicals sometimes harmful to humans or other organisms.[6]

That's the key dynamic that describes the speed of human development, and consequently its effect on the Earth's flow of carbon. We communicate symbolically, voluminously, quickly, outside our bodies, at will, and in a realm parallel to biochemistry: the "human mental world."[7] Before language, thought didn't travel very far at all, at least not with any specificity. And before communities speaking different tongues deciphered each other, ideas remained rooted in linguistic communities. The speed of thought has accelerated throughout human and cultural history, first as language developed, and then as transportation and communication systems allowed people to share ideas with each other faster and faster. The horse, train, airplane, and Internet have dramatically reduced the effort required for individuals to communicate. Biochemical communication is still rooted in molecules, with physical and chemical limits on where, how far, and how fast its information can travel. Ideas can fly around the world off the tip of the tongue or the push of a button or await an interlocutor hundreds of years in the future. Civilization rose from sophisticated communication but primitive tools.

It's hard to think of anything we do that does not correspond to what some other organism does on some level. Skyscrapers didn't appear regularly in nature until the Woolworth Building went up on Broadway in Lower Manhattan in 1913; yet termite mounds loom proportionally as high as skyscrapers. Humans have no monopoly on communication. Ornithologists have transcribed birdsong for decades. Vervet monkeys wield a vocabulary of a dozen or so words.[8] Humans just do things on scales of time and space previously unachieved. Our influence over the planet is unique in its degree in both rate and magnitude.

Historians of economics and technology might not have come up with their own Central Dogma. Their rigor has deepened the analogy expressed in Dawkins's memes by grafting the Central Dogma onto patents, manufacturing, and products. Patents are a kind of "genetic code" of the inventions put forth within them. Ribosomes, which are largely RNA machines, are referred to as cellular "protein factories." It's easy to flip that simile around and think of factories as expressing design information into the industrial equivalent of a phenotype, an organism's observable characteristics. Take car making. Takahiro Fujimoto, a University of Tokyo professor and auto industry expert, belongs to a school of enterprise specialists who see explanatory power in applying the DNA-RNA-protein dogma to the factory. Teams of individuals come up with a company's proprietary design information—its genetic material. Workers express it, through organization, factory layout, equipment, and task assignment, into a finished product. Fujimoto views products as "media infused with information," or "packaged information."[9]

"The industry of industries"—Peter Drucker's phrase for car making—is an acceleration of a pattern that emerged long ago in human evolution. Everything humans make is either mostly carbon or forged by its flames. From hearth to Honda, fire has always forged complexity in human social interactions and accelerated innovation. Arizona State University professor Stephen Pyne calls fire the "paradigm for all of humanity's interactions with nature." Industrial combustion reverses photosynthesis, turning O_2 and ossified carbon back into CO_2 orders of magnitude faster than oxygen-breathing organisms could unaided by humans' machines.

Fire is incandescent carbon, freed from bonds to other carbon and hydrogen, and smashing into other molecules and fragments billions of times a second. Think about what happens if you pass your finger quickly through a glowing flame. Your finger will have a little greasy black film, which is carbon freed from the wax, but depleted of the energy that made it glow. A candle flame is a storm of collisions, unromantically described by combustion scientists as a "spatially localized self-sustaining, chemical reaction zone."[10]

Fire took time to ignite. Oxygen was nearly absent from the atmosphere until 2.32 billion years ago, so the first half of Earth history was fire free. Combustion must have been confined to its slow, controlled form, respiration, trapped within the membranes of microorganisms. The rise of oxygen may not have literally set the world on fire, but its accumulation caused the Earth's atmospheric methane and ethane to combust, dispatching perhaps 10°C worth of heat off the planet and freezing the Earth into a snowball.[11]

Kindling took almost another 2 billion years to emerge after that. Tall trees sprouted 400 million years ago, offering carbon- and hydrogen-rich lignin and cellulose as fuel for oxygen to attack. Whether in Ferraris or forests, molecules need a catalyst from a spark plug, lightning, or just some plain old heat source—to ignite. Friction from rubbing two sticks together sometimes suffices. The earliest known natural forest fires date to the Devonian period, which ended 359 million years ago.[12]

When hominins first controlled fire is tough to pin down. Our ancestors may have ventured about Africa and into Eurasia more than a million years ago. They would have relied on fire as a "portable climate," and for cooking. Unfortunately, most charcoal ash disappears before sediment buries and preserves it; water washes it from caves, and arid equatorial climates oxidize it. Charcoal sites dating to 1.5 million years ago are of disputable hominin origin.[13] Early humans would have needed fire in Europe or northern China perhaps 500,000 years ago, but clear-cut hearths that old have not turned up. The surest hominin fireplaces date from 250,000 years ago.[14] Conventional wisdom holds that hominins controlled fire by some of these earlier dates, even though archeological evidence is either equivocal or missing.

Frank Niele, a senior research scientist at the Royal Dutch/Shell laboratories, sees the control of fire as the advent of human "societal metabolism," which he defines as humans' ability to funnel energy toward and shape matter into needs dictated by survival, social interaction, and teaching.[15] Fire, tool use, community structure, and brain development leapfrog each other down the ages. Hominin crania grew at an accelerating clip through the periods of hunting, gathering, and scavenging. Over the 6 million years or so of prehuman history, the brain has grown three times its original size. Human brains are large to begin with—six times bigger than scientists would expect by comparison with other mammalian brain-body ratios.[16] Fossils and archeological discoveries suggest that social development, tools, and brain growth transformed each other with increasing momentum. Human brains stopped getting larger 100,000 to 50,000 years ago.

In modernity we focus not on brain growth but on another result of social interaction and toolmaking: economic growth. The hearths of Europe and East Asia prepared much more than food. They fired copper and tin into bronze, and iron and black carbon into steel. Virtually everything is either made of or forged in carbon, or more likely, both. Resources, knowledge, and tools begat more resources, knowledge, and tools. By the mid-nineteenth century, for reasons debated in many books— significantly Jared Diamond's *Guns, Germs and Steel*—inventors in France and Germany stood poised to exploit their resources, knowledge, and tools in a way that would transform the Earth's surface and atmosphere in unfathomable ways.

Lift open the hood of a car today and you will see structures and systems that date back decades, or in some cases more than a century. The body itself is a legacy of the horse-drawn carriage, the width that can accommodate two or three people sitting next to each other comfortably. Cars and horses coexisted on city streets for several years at the turn of the twentieth century. The need for change was both in the air and along the curb. Horses were nuisances. They polluted cities with manure and seemed to have minds of their own.[17]

A good year to begin a chronology of the automobile's invention is 1860. In that year, Frenchman Etienne Lenoir built a stationary internal combustion engine that burned gas derived from coal. He put an engine on wheels two years later. The contraption ran, though inefficiently because he had not discovered that air and fuel perform better if mixed and compressed before they burn.[18]

French civil servant Alphonse Beau de Rochas patented a system for turning hydrocarbon fuel into mechanical motion. He approached Lenoir about funding and support, but was turned down. German inventor Nikolaus Otto independently invented the same system several years later. Lenoir, later realizing he dropped the ball on what turned out to be the greatest breakthrough since mitochondria, began producing horseless carriages himself, based on the Beau de Rochas patent. Ignorant of the patent, Otto sued and lost. He is nonetheless accorded historical credit for inventing and building the first four-stroke internal combustion engine.[19]

Lenoir, Beau de Rochas, and Otto were not working in a vacuum. Many people were working on the transportation problem, and had been for a century. Electric cars enjoyed limited success into the early twentieth century, despite the bulk and unreliability of their batteries, and an occasional tendency to explode. The hydrocarbon-fueled car was a logical progression from the steam car, which burned carbon that heated water into the steam that powered mechanical movement. The gasoline engine's inventors realized they should take out the middleman, the steam, and just burn hydrocarbons into mechanical energy. Gasoline was a throwaway by-product of kerosene refining until the early 1900s, used sometimes in solvents or as fuel for stoves. In 1892, two cents a gallon was a decent price. For another thirty years, apothecaries were the makeshift filling stations.[20]

The French, Germans, and Italians ran a decade ahead of the United States in automobile innovation and production through the 1890s. Karl Friedrich Benz put an internal combustion engine on three wheels in 1885, after a decade of selling stationary engines. Benz's car used an electric spark igniter, more efficient than previous flame igniters. He patented a radiator the next year, the automotive equivalent of hairless sweating. Gottlieb Daimler, who had worked for Otto, built a 900-rpm motorcycle in 1885. Inventions thereafter shot forward. John Boyd Dunlop patented pneumatic tires that made it onto cars by 1897. Benz started making the gearbox in 1899, which allowed the engine to perform better at higher speeds. The self-starting ignition came about in 1911, three years after the Model T itself.[21] Few, if any, truly revolutionary changes in automobile technology have come along since the end of the nineteenth century, with the exception of the microprocessor. All cars have become electric vehicles in a sense. Today, cars demand fuel for much more than mobility: power windows, air-conditioning, DVD players. These features now use about as much energy as the car itself, in part due to progress in engine efficiency. In the 1920s, the ratio of fuel used for mobility versus other things, mainly sparking, was 9:1. Today the ratio is closer to 1:1.

Car making started as a craft performed in small shops, each drawing parts haphazardly and honing them as needed. Evelyn Henry Ellis, a member of parliament, became the United Kingdom's first-ever driver in 1895.[22] The previous year, he commissioned the French company Panhard and Levessor (P&L) to build him a car. Every car back then was a unique work of artisanship, even if two vehicles shared the same blueprints. There were no standardized parts, and consequently no mass-produced models. P&L had to make individual parts fit individual cars. The company would receive a batch of parts and file one piece down to a useful size and shape. The next piece needed filing to the size of the previous one, and so it went all the way through production. Each P&L car effectively had its own standard-size parts, determined as needed. Imagine going into a hardware superstore today that sold only ingots of various sizes and a huge array of files with which to grind them into screws, nuts, and bolts. Manufacturers set their engine on a chassis, covered it with a hood and set it alight. The word "chassis" is a nod to the French influence on the commercialization of automobiles.

Ellis took liberties on his first road trip, doubling the 4 mph speed limit for horseless vehicles. Within a year, cars challenged the new limit of 12 mph. Soon, shops all over western Europe and North America assembled cars, in the range of dozens to hundreds a year.[23]

The beginning of the twentieth century brought new machines capable of shaping hardened steel parts. Henry Ford saw the possibility this posed. If parts came standard, his factory could eliminate the expensive and time-consuming work of filing parts to fit each other. This development, not the assembly line, prefaced the explosion of Model Ts on the world's roads. Interchangeable parts enabled mass-production to churn out cars. Ford engineers designed an incredible, three-dimensional mechanical jigsaw puzzle, so easy to assemble that every owner on his own could fix it or scrape the carbon buildup out of the cylinder and off the piston with a putty knife.[24]

To produce cars, humans had to reinvent the notion of standard parts and systems, something life has "known" since its inception. The combined innovations of standard parts and the assembly line produced astonishing gains in productivity. In the fall of 1913, craft-style assembly required about 750 person-minutes to assemble a vehicle. By the spring of 1914 it took Ford workers 93 minutes.[25] The Ford boom had begun, leading to inexpensive cars, a doubling of Ford wages (to $5 a day, or $100 in current dollars), and a nation of committed drivers. Authors of a U.S. Energy Department report on research needs in the automotive sector introduced their study this way: "Our Founding Fathers may not have foreseen freedom of movement as an inalienable right, but Americans now view it as such."[26]

Toyoda Automatic Loom Works opened as a manufacturer of textile industry equipment in the 1880s.[27] In the early 1930s, Kiichiro Toyoda, the founder's son, and a team of engineers started taking apart Fords and Chevys to gain insight into U.S. automobile manufacturing. The project

*Henry Ford edged out Alexander Winton in a 1901 race in Grosse Point,
Michigan—at the beginning of the carbon age.*

translated into Toyoda's first vehicles, trucks, which worked passably well
despite the company's reliance on hand tools rather than expensive Ford-
style stamping machines. The company made modest progress and in
1937 spun off Toyota Motor Co. After World War II, Toyota emerged ambitious but struggling.

Kiichiro Toyoda's goal to reach U.S. production levels within three
years after the war's conclusion proved far too ambitious. In 1950—five
years before Detroit hit its own peak production—the company put out
only thirty thousand vehicles, below its prewar numbers. A dire postwar
economic environment, exacerbated by credit restrictions imposed by the
U.S. occupation, nearly put the company under. Toyoda's move to fire two
thousand people—a quarter of the workforce—led to a strike, and his res-
ignation. His son Eiji and chief engineer Taiichi Ohno put the company
back on its feet and gave the automotive industry its current winning
manufacturing model. Visionary leadership and planning helped. But
Toyota Motor Co. became the world's second-largest carmaker mostly
through desperation, and myriad tweaks in production that overall created a masterfully efficient organization. Ohno toured a U.S. carmaker in

1956. In the decade since the war ended, Toyota had already squeezed tenfold productivity improvements from factories—making it a smoother operation than the one Ohno visited.

Design information is stamped into steel in automobiles, the industrial equivalent of RNA stamping out proteins. A sheet of steel slides between two dies, one the negative shape of the other. Then thousands of pounds of pressure warp the steel into a fender, door, or trunk. In the early 1950s, Toyota did not have the capital Detroit had to stock its factories. Nor could it afford losing days of production to switching the dies whenever production changed, as the industry leaders did across the Pacific. Necessity mothered invention. A rotating die system helped Ohno shrink the die-changing process from a day to two or three minutes, an early victory for what would later be heralded as the Toyota Production System. Austerity-enforced inventiveness paid off early, even if Toyota needed another two decades to become a world power.

Another innovation teases out the similarity and differences between how people and biochemistry each make things. Cells are so complex and effective because the whole edifice relies on individual nodes of contact between molecules. They have to swim through the proper channels and fit together properly or they won't work. Essentially, Ohno magnified by many times the number of nodes of contact between assembly-line workers and the product, and among each other. The net effect was that more eyes and hands and brains made the system incrementally more robust, gradually rooting out errors. Ohno hung cords above every workstation, empowering every member of the manufacturing team to shut down the assembly line whenever he spotted a mistake. When someone pulled the cord, his colleagues would gather around, determine what went wrong, at what stage of the process, and solve the problem at the source. Errors regularly shut down the lines early on. But then something remarkable happened. Errors diminished. Responsibility for quality became decentralized. End-of-the-line fix-it specialists became as useful as the human appendix.[28] Detroit automakers maintained inspectors at the end of the production line who fixed imperfections in the car. Without planning it, Toyota "evolved" a system that erred less.

Car buyers rewarded these innovations. Toyota sold more than 200,000 cars a year through the 1960s, before export markets started to open. They exported 40,000 cars in 1960, a million in 1970, 6 million a decade later, and 7 million in 2000.[29] In the first quarter of 2007, Toyota was the world's largest carmaker, before dropping back to second place. Having come late to the hybrid game, Ford began in 2004 to license Toyota's hybrid gasoline-electric technology, in exchange for licenses on both diesel and direct-injection engine patents.[30]

Regardless of their manufacture, gasoline automobile engines require a slow, smooth-burning hydrocarbon to maximize energy efficiency and reduce pollutants and engine knock. Spark-ignition engines (as opposed to diesel) favor branched alkanes, or chains of singly bonded carbons with hydrogen in tow. Iso-octane is a branched alkane, and its dense molecular structure makes combustion proceed more slowly than its straight-chain sibling, octane. The molecule's shape also protects it from free radical attacks, the same OH− and H+ ions that wreak havoc on our cells as we age.[31]

Internal combustion is chemically no different than burning a candle—just more efficient. At ignition, oxygen swooshes in from the room and reacts with carbon and hydrogen around the hot base of the flame. Ideally, hydrocarbon burning should yield CO_2 and H_2O, but wax vaporizes incompletely. Instead, oxygen is depleted in the center of the flame, carbon atoms cluster together into molecular carbon bits, and fly off, ready to assemble themselves into larger structures thousands of atoms in size, or more. Along the perimeter, atoms coalesce into new molecules every picosecond. Oxygen knocks off a hydrogen atom here or there to create free radicals. Radicals beget more radicals and leave carbon atoms behind to clump together. When they can, the radicals end up forming water vapor and zipping away. Big soot particles condense and waft toward the ceiling. There's also leftover carbon in that greasy black stuff and molecules aloft that are too small to see.[32]

Gasoline engines are more efficient. Fuel vapor mixes with air before they burn together. This reduces soot and the exotic carbon compounds forged in flame. Combustion in automobiles is also much more complicated than a chemical hop, skip, and jump from hydrocarbon to CO_2 and H_2O. In fact, it's so complicated that fuel chemists can only estimate the number of reactions involved. Methane is the simplest hydrocarbon, one carbon with hydrogen atoms bonded to it, positioned as if they stood at the vertices of a tetrahedron. Burning methane might produce 100 different intermediary molecules and 250 chemical reactions. Heptane, a component of gasoline, and iso-octane can break down into 1,000 different molecules and fan out into 8,000 reactions.[33]

Imperfect combustion becomes a public health and environmental problem when it builds up as pollution. In the early 1970s, governments mandated that automobiles be equipped with catalytic converters, which finishes off some molecules in the exhaust stream that burned incompletely. Gases on their way out of the car pass through tubes in the converter. They seep behind ceramic lining toward platinum and rhodium catalysts, which burn up imperfectly combusted particles. The rhodium turns nitric oxide back into oxygen and nitrogen. The platinum catalyzes oxygen atoms to react with CO, making CO_2.[34] Honda avoided the catalytic converter altogether by introducing its Compound Vortex Controlled Combustion in the early 1970s. It controls the amount of air and fuel mixed, and the speed with which they burn. This technology avoided the need for an emissions scrubber because it caused a thorough burn on the front end.

Fossil fuel–burning automobiles and electricity generators are ultimately solar powered. Sunlight welds airborne carbon dioxide into plant sugars inside a leaf or algae or other photosynthetic material. Most fossil fuels are just plants and algae with the water and oxygen pressed and baked out of them. Jeff Dukes of the University of Utah has estimated that 98 tons of vegetation over hundreds of millions of years yield 1 gallon of gasoline.[35]

It would be nice to speed up that process, or more likely, deploy alternatives. Sugar cane, switchgrass, French fry oil, and turkey offal can or do fuel internal combustion engines. Avant-garde energy research challenges the divide between what is natural and unnatural even further. Fuel chemists have refined diesel gasoline from "salmon oil with high content of free fatty acids." The decline of wild salmon makes aqua-gas hard to envision. Even if the salmon weren't overfished, the resources to feed, raise, and refine enough fish oil to make fuel make this project rougher than swimming upstream.[36] Even more unusual possibilities exist. The International Energy Agency has catalogued the use of "animal tallow derived from specified risk materials, dead stock and downer animals as feedstock for the production of biodiesel."[37] This process would refine mad cow–afflicted bovines into gasoline.

If scientists can get the chemistry right, and engineers can build the refineries, machines will never know the difference between traditional petroleum gasoline and biofuels.

Evolution and human economic activity are both manifestations of complex adaptive systems. A scholarly branch of economics, called evolutionary economics, doesn't try to explain human social and commercial interactions in terms of biology. But it does compare evolution and economics, and teases out the abstract similarities that tie them together—replication and selection to name the two most important.[38] Guided by the general framework of imperfect replication and selection, more explicit analogies between the "natural" and "unnatural" modi operandi might be worth a look.

Ford Motors' early success came from bulletlike ambition to feed insatiable demand. One might see broad evolutionary patterns between the cyanobacteria's success and Ford Motors', and the coccolithophores and Toyota. Cyanobacteria did something no organism had done before. They plastered carbon into their cells harnessing electrons taken from water. The abundance of raw material (carbon and water) and this ability to bring them together allowed them to live in as many places as they do, in such abundance. By contrast, coccolithophores thrive in hardship, in waters so devoid of nutrients, such as phosphorous and nitrogen, that they must build their shells through innovative biochemical shortcuts. Polysaccharides—molecules of only C, H, and O—guide shell production. In other organisms, proteins build shells. The coccolithophores just can't afford it. Innovation, evolutionary or technological, can emerge from either abundance or deprivation.

Five characteristics of evolution were discussed in chapter 2:

- Biomolecules and the cell's environment interact with DNA to grow an organism and keep it alive.

- Populations evolve, not individuals.

- Evolution runs forward in many directions at once, like the branches of a tree.

- But it does so without any goal or trajectory other than survival amid continuously changing conditions.

- Natural selection determines which adaptations are beneficial.

Only the last of these statements appears to apply unaltered to human technology. Novelty comes from imperfect reproduction and selection. The first statement doesn't apply because human communication is symbolic, not electrochemical; car executives don't make cars by rubbing official patent documents into steel. Second, companies don't make populations of products on any scale comparable to nature. They generally design and make individual products, and hone them into "new and improved" versions. Compare the number of cars that a company designs annually, to the 30 million eggs laid every year by an ocean sunfish. Third, economies look more like upside-down trees; entrepreneurialism runs strong in the United States, but economic niches seem to often trend toward monoculture, be they in operating system software, telephone service, or media. Fourth, the whole notion of progress, however defined, separates industry from adaptation to change, a hallmark of evolution.

One-time breakthroughs honed by powers of incremental alteration and selection have driven human development throughout the ages and around the world—literally. So close is the natural selection analogy that John Maynard Smith and Eörs Szathmáry explained in their book, *The Origins of Life,* that the motorcycle, "if you like that kind of thing," works in a manner reminiscent of endosymbiosis, the merger and acquisition of one cell by another that wrought eukaryotes with their mitochondria and other organelles. Automobiles emerged from the marriage of internal combustion engines and bicycles. This industrial endosymbiosis gave rise to the world's automobile population—in a geological blink.[39]

8 | The Physical Restraint on Fantasy
The Art in Carbon Science

> Magic is antiphysics, so it can't really exist. But it shares
> one thing with science. I can explain the principle behind
> a good science experiment in 15 seconds; the same with
> magic.
>
> —Koji Nakanishi

Plants, microbes, and animals benefit from millions of years of evolutionary experiments on carbon molecules. They produce an array of small molecules for use as everything from poison to painkiller. Sometimes they make these complex structures for no good reason that scientists can immediately discern. Chemists take advantage of them anyway. Between 1981 and 2002, 74 percent of new anticancer drugs and 78 percent of antibiotics were natural products, derived from them or inspired by them.[1] Even when it doesn't result in a new drug, studying naturally occurring small molecules reveals new tools for further discovery. And often the "total synthesis" of an interesting molecule is a feat in itself, as literal a victory of mind over matter as they come.

Ginkgo biloba's ancient use in medicine mimics its deep evolutionary history. The potentially 5,000-year-old Chinese *Materia Medica* describes *Ginkgo* as an aid to blood circulation and lungs. It acquired ever-more folk uses over the centuries, from an antidementia agent to a remover of freckles. Since the 1980s, it has been known as an anti-platelet aggregating factor (anti-PAF), which means it prevents blood from clotting. Research continues into its possible ameliorating effects at the onset of Alzheimer's disease.

Perhaps its perceived medicinal benefits, the tree's beauty, or both bestowed sacredness upon the tree in parts of the Far East, particularly in Japan, where the tree was named. A Japanese word for ginkgo, *icho*, turns up in culture and language. Living and inanimate things have *icho* as a descriptive element in their names, from crabs and the *icho-ha-kujira* (ginkgo-tooth whale) to traditional musical instruments and the hair bun of junior sumo wrestlers. *Ginkyo* became "ginkgo" when a Dutch explorer brought *Ginkgo* to Europe (for the first time in 2 million years).[2] The Swede Carl Linnaeus, inventor of biological nomenclature, picked up on that transliteration, and it stuck.

In 1963, a typhoon ravaged Sendai, Japan, destroying homes and uprooting trees—including many *Ginkgo biloba*. The damage was an ironic boon to research chemists at nearby Tohoku University. It is not a figure of speech to say that ginkgo's health benefits are rooted in the East. The active molecules occur in great abundance in the bark that gloves *Ginkgo* root networks. The scientists received permission from the Sendai City Office to haul away about 220 pounds of otherwise protected *Ginkgo* roots, cut from felled 5-foot thick trees.[3] In 1963, no one knew precisely what these molecules looked like. By 1967, they did.

Plants make gazillions of extremely complex molecules with their spare energy and materials. Part of the genius of trees is that they find useful ways to store material that is secondary to staying alive. They have ample opportunity, material, and energy to design new molecules to tinker with, to offer animals as signal or food—or poison—or neither. These molecules are not modular, like cars or televisions. They were never "meant" to be reverse engineered. Yet making these molecules unnaturally, in labs, is close to what scientists must do. They figure out ways to simplify molecules, piece by piece, then put them back together.

Ginkgo makes particularly interesting chemicals, "bitter principles" that scientists identified for the first time in the early 1930s (in part by tasting lab samples). Another generation passed before chemists could look with greater precision, taking advantage of powerful tools that peer into the molecular world. Louis Pasteur, the founder of microbiology, famously quipped that chance favors the prepared mind. More often than not, it also favors the operators of fancy lab equipment.

At first the Tohoku chemists also deployed the least expensive lab apparatus imaginable—their tongues—to find the target chemicals. Eventually, their taste buds could no longer guide them through the acrid chemicals. Extractions, measurement, and tedium only the obsessed can abide led to the further isolation of their four target molecules: ginkgolides A, B, C, and M. Separating the ginkgolides from each other proved a formidable challenge. Instead of crystallizing into four separate compounds, the molecules crystallized together. After ten to fifteen arduous steps, the ginkgolides A, B, C, and M molecules finally segregated themselves under physical duress, into gram amounts.[4]

With the molecules purified, the real work began. What did they look like? Nuclear magnetic resonance imaging is the same technology that doctors use to see inside the body. (Medical practitioners renamed the technology MRI, magnetic resonance imaging, to prevent the word "nuclear" from scaring away patients.) Through a combination of NMR studies, x-rays, and traditional chemical techniques the ginkgolides showed up, beautiful sculptures of six rings of five atoms each, tightly wedged together in a network of bonds that gives the ginkgolides their remarkable stability.

The ginkgolides are among the most complex molecules known of their size. The work isolating the ginkgolides was intense, the hours long, and the result historic. Technology played a key role in the ginkgolide structural studies, but some scientists feel new machines took some of the fun out of the process. "For me, it was the last classical and romantic structural study," recalls Koji Nakanishi, who shortly afterward began a distinguished career at Columbia University in New York studying how nature makes complex molecules.[5] Chemists also know him for his avocation. Nakanishi is a magician.

The synthesis of ginkgolide B became a sword in the stone, a natural target for chemists to rebuild by unnatural means, from scratch, all twenty carbon, twenty-four hydrogen, and ten oxygen atoms in their appropriate places. Even among complicated molecules, the ginkgolides are complicated.

Significant research has been conducted on the ginkgolides' potential health benefits, particularly that of ginkgolide B. The first concrete discovery wasn't made until 1985 and much of it is ongoing.[6] The ginkgolide structures were extremely taxing to discover yet much easier than figuring out what the body does with them. In 1967, the Japanese scientists looked at the compact, complex molecules and decided its medicinal mystery would remain locked inside them for some time.[7]

Half a world away, at Harvard University, chemical synthesis was emerging as a mature field—a fusion of art and logic—tempered by the realities of what atoms will and will not do. The analytical habits created at Harvard, Oxford, the Swiss Federal Institute of Technology at Zurich, the University of Illinois, and elsewhere created the scientific habits that led to the discovery of many modern pharmaceuticals.

Labs are structured as intellectual family fiefdoms. A professor "raises" his graduate students, who grow up and fan out across the world of research universities and private industry. Virtually everyone's intellectual ancestors can be traced back to J. J. Berzelius, the Swedish chemist who first called carbon "C." Every generation tends the repository of knowledge, weeding out its predecessor's bad ideas, answering some of their questions, and asking many of their own.

"The transformation of molecules into new molecules" is a simple description of chemistry. In synthetic chemistry, scientists make natural products via unnatural routes. Synthetic diamonds are real diamonds, only made in labs from natural gas plasma or in a heat-and-pressure apparatus that mimics and accelerates the Earth's interior diamond oven. Synthetic rubber covered tires of army Jeeps during World War II, and synthetic gasoline fueled their tanks. Both were made from coal. These products are identical to "natural" ones but are brought into existence by laboratory experiments conducted on petroleum by-products.

Synthetic chemistry lunges forward by discovering new ways to cook molecules. "Chemical synthesis is uniquely positioned at the heart of chemistry, the central science, and its impact on our lives and society is all pervasive," E. J. Corey of Harvard has said. Academic chemists are in the business of learning new pathways for turning specific molecules into other specific molecules. Synthetic chemistry fuses the intent of the artist with the fastidiousness of atoms. Imagine if Rembrandt had to contend with paints that interacted with each other violently, that underwent what one chemist calls "unusual and bizarre rearrangements." What if a dab of brown at the bottom causes the red on the upper right of the canvas to jump to the upper left or turn blue, or turns the portrait-subject's ears upside down or replaces eyes with golf balls? What if a single brushstroke of white in the wrong place causes the canvas to unravel into a single fiber? Yet that's what synthetic chemists have to contend with. They sculpt molecules with atoms. And atoms have their own ideas about the art of the possible—what the late Robert Burns Woodward, possibly the king of synthetic chemistry, called "the physical restraint on fantasy."[8] Add a bromine atom here and the −OH group on the other side of the molecule snaps off. Sometimes chemists must add protective groups, like putting masking tape over one half of a canvas so you can paint the other half. Like painting, in chemistry, it's not always so easy to get the tape off.

Today, synthetic chemistry is the repository of two hundred years of aggregated knowledge. Experiment by experiment, successes and more frequent failures, chemists amassed libraries of reactions and the catalysts that speed them up. Friedrich Wöhler was the first scientist to make a natural substance by unnatural means. He synthesized urea in 1828, the key ingredient in urine. ("I have to let out," he wrote to his mentor, "that I can make urea without needing a kidney whether of man or dog."[9]) His historic success was tempered even in his own mind by the primitiveness of the science, which he described thusly, "Organic chemistry nowadays almost drives me mad. To me it appears like a primeval, tropical forest full of the most remarkable things, a dreadful endless jungle into which one does not dare enter for there seems to be no way out."[10]

Chemists in the nineteenth century amassed practical tools for transforming molecules long before they knew what they were doing or even what atoms were. The discovery of protons, neutrons, and electrons was a long way away, let alone the discovery of quarks, which make up protons and neutrons. Initially, atoms were accounting mechanisms, tools for explaining the composition of molecules but devoid of meaning themselves. They were placeholder concepts in the way that "consciousness" or "dark energy" are placeholder ideas today for unexplained phenomena.

Chemistry took off around the middle of the nineteenth century when scientists began to think about the structure of molecules they made in their labs. They developed so-called structural theory, a way to represent on paper the physical relationships atoms had to each other. They knew that carbon bonded with up to four other atoms at the same time. In the early 1870s, Jacobus Henricus van 't Hoff proposed that the carbon atom's preferred bonding structure is three-dimensional—as if the carbon sat in the middle of an imaginary tetrahedron, bonded to other atoms at the four vertices.[11] To convince colleagues, he fashioned wooden tetrahedrons and dispatched them to scientists across Europe.

University and industrial scientists spent decades discovering reactions, cataloguing them, and shuffling them together into books. Patterns emerged that allowed them to predict and explain how classes of molecules behave along general principals. The knowledge accrued and empowered chemists to create materials never before occurring in the known Universe. In 1881, Friedrich Konrad Beilstein started listing the known carbon compounds. His initial tome became a three-volume second edition in 1889. The fourth edition, which came out thirty-one years after his death, filled twenty-seven books. Now it's a proprietary online database expanding toward infinity.

The fundamental goal of synthetic chemistry is to build molecules as complicated as possible in the minimal number of steps. Chemists draw on the accumulating, historical reservoir of reactions, reagents, and catalysts that en masse explain how the molecular world (and thus the world) works. Research scientists can choose to mimic natural products or create new, often beautiful, symmetrical molecules that may never have existed before. They can search for more effective ways to build known pharmaceuticals. They "improve" natural molecules for use as drug treatments. Or they can pursue basic science research that continues the hundreds-year-old chemistry enterprise of watching atoms and molecules behave or misbehave. They look for surprising behavior and then try to explain it. That's what drives the field forward.

Synthetic chemists decide what chemical targets to choose based on many criteria. Sometimes scientists choose molecules because they have a hungry pharmaceutical industry and suffering humanity waiting for them (not to mention research grants). There are many famous examples of syntheses. The yew tree was a known poison for at least two millennia before researchers discovered at the prompting of the National Cancer Institute an extract that killed leukemia cells and inhibited the growth of other tumors. They called the active molecule Taxol, but couldn't do much more. To gather enough Taxol to give one patient one dose, about 300 mg, scientists would have to uproot and gut an entire century-old tree. The Pacific yew is a late bloomer. Synthetic chemists eventually built the molecule independently, five times on five different paths.[12] This work led to understanding of the molecule and its eventual manufacture through a fermentation process.

Viagra, Prozac, Lipitor—some of the most famous molecules of our time—all have fascinating stories of serendipitous discovery behind them. Today, the search for new biologically active molecules has cross-pollinated with mass production. Robots can test thousands of molecules using high-throughput screening, a way to find potential drugs that interact with genes. Chemists can take genes they are interested in and splice them to luciferase—the protein that causes fireflies to light up. When a research molecule activates a gene, it flashes.

At the other end of the spectrum are the purely scientific exercises, target molecules that beckon like Mt. Everest. Every new synthesis expands chemical knowledge and adds arrows—reactions—to the quiver for future generations. Toward the latter end of the spectrum falls ginkgolide B.

Neanderthals knew specific plants relieved some ailments. Hippocrates wrote 2,500 years ago that women in labor appeared to be comforted by ingesting bits of bark from the willow tree. Pliny the Elder, writing several years before he and Pompeii died in Mt. Vesuvius's ash, chronicled the benefits of willow bark for relieving pain and treating warts. Using willow to treat warts is a practice that appears to have died with Pliny the Elder, and Pompeii.

Investigation into what properties made willow work ultimately had to await the decline and fall of the Roman Empire, the Dark Ages, the Renaissance, and the Enlightenment. The first known query occurred in 1763, when the Reverend Edward Stone wrote to the president of the Royal Institution about the properties of willow. He undertook the first studies of what decades thereafter would be called salicytic acid. By 1827, chemists were able to extract a carbon-ring-based molecule called salicin. This is metabolized into salicylic acid—willow bark's active ingredient.

Most nineteenth-century discovery came from felicitous tinkering or outright accident, as in the case of Friedrich Wöhler's synthesis of urea. Chemists picked available starter molecules and hung other molecular groups from them, building them up and changing their properties.

Hermann Kolbe was one of Germany's great mid-nineteenth-century organic chemists, and the first to make salicylic acid from scratch. In fact, this feat was the first chemical synthesis to be called such. Kolbe heated sodium phenoxide—essentially a hefty dose of sore throat spray—with carbon dioxide, under pressure. Out came sodium salicylate. A student of Kolbe scaled that lab experiment to a commercial scale, and sold the drug to hospitals for use as a fever reducer and pain reliever. To everyone's chagrin, it upset stomachs. The product was perfected by a dye chemist for Bayer, which patented acetylsalicylic acid as aspirin.[13]

Aspirin gained renown as a fever reducer and pain killer and later was realized to have anti-inflammatory properties, too. It is an example of an altered natural chemical with enormous commercial value. Aspirin is also an anti-platelet-aggregating factor, which is why doctors sometimes recommend it for patients with a history of heart disease. Ginkgolide B was discovered in the mid-1980s to be an anti-PAF. It's important to note that the Food and Drug Administration does not regulate ginkgo biloba, and most Western doctors do not suggest heart patients take it the way they do aspirin. Much more is known about the latter. Medical practitioners do warn against taking aspirin and ginkgo biloba supplements together, however, since they are both known anti-PAFs.

Every science seems to undergo an age of giants, when the field is new and an entire body of knowledge is consumable by one person. Then the science matures, scientists' numbers grow, and they specialize, fractal-like, into ever-branching subfields. Chemical synthesis became an endeavor for giants of the field in the first decade of the twentieth century. Molecules were captured and paraded about like the biggest fish of the year or the largest set of antlers.

The first third of the twentieth century saw theories emerge to explain or predict the behavior of atoms. Physicists by the 1920s had described the quantum world. Wolfgang Pauli's exclusion principle, which states that electrons refuse to share each other's atomic territory, or orbitals, is considered the demarcation point between physics and chemistry. Pauli himself anthropomorphized electrons as "antisocial" because they will not admit others into or past their orbitals, each characterized by a specific set of quantum-world factors. Chemistry is a game of electrons, which conventionally makes the exclusion principle a line in the sand.[14]

Within three years of that discovery, the American Linus Pauling took a monumental first stab at explaining how behavior in this quantum world results in the tetravalent bonding of carbon.

By World War I, scientists had amassed a body of observations about how atoms and molecules behave and saw sharper patterns within them. Patterns emerged in the catalogue of reactions that gave order to the dense jungle described by Wöhler a century earlier. This wider classification of how chemistry works yielded to theoretical stirrings that gave even deeper meaning to why atoms and molecules should behave in the way that they do.[15] Theoretical chemists learned that bonding is a game of electrons. Covalent bonding occurs when unpaired electrons find shell mates in the outer limits of nearby atoms. Scientists also discovered the third "C" of C science. "Constitution" is the basic architecture of a molecule. "Configuration" refers to either of two mirror-image arrangements four atoms can take when bonded to a single carbon. Chemists informally call this right- or left-handedness. "Conformation" brought understanding of molecular flexibility—their "floppiness" and the ability of a molecule's atoms and functional groups to find a comfortable place.[16] This discovery helped chemists predict how molecules might behave in new reactions.

The 1950s and early 1960s was a time of rapid transition and intellectual giants. Industry drew on chemists' research and guided them with the demands of medicine. Academia and the pharmaceutical industry became entwined in a delicate dance, close but not too close, that continues to the present.

In 1960, Robert Burns Woodward's lab at Harvard became the first entity in three billion years to create a pathway that yields chlorophyll *a*, a pigment that cyanobacteria use to capture energy from light and that plants exploited once they abducted the cyanobacteria. Woodward saw that chlorophyll is a symmetrical, cloverleaf molecule filled with highly substituted carbons—that is, filled with carbon atoms each bonded to four nonhydrogen atoms. Instead of thinking about these arrangements as a roadblock, he saw that their formation was the way to guide the skeleton into place. Woodward's chlorophyll began with a common, off-the-shelf chemical, which was quartered and then welded into the chlorophyll skeleton. The last steps involved sprinkling the molecule with its outlying molecular groups. As in all of the landmark syntheses, Woodward used master strokes, including a novel light-induced reaction and cunning, final isolation of the desired molecule from its chiral or mirror image. The effort took four years.[17] The synthesis needed fifty-five steps and seventeen post-docs. Every reaction in a synthesis carries a theoretical yield rate. But the experimental yield always falls short of the theoretical yield. That's why the shorter a synthesis, and the higher the yield at each step, the more of the targeted material will be produced. Many details of Woodward's chlorophyll synthesis were never formally reported.[18]

Woodward's chlorophyll a is a wonderful index of the simultaneous greatness and limitation of the scientific enterprise. The work was feted in 1960, when *Time* magazine made Woodward and fourteen other U.S. scientists its Men of the Year. On the other hand, from a biological perspective, four years of work yielded only chlorophyll, not the suite of fatty acids that fasten it to a cell, nor the complicated chain of proteins that carry the light-excited electrons to the cell's battery molecules—let alone putting all these things together in a living cell. Plants not only make chlorophyll but stack it and wire it into the photosynthetic membrane. Woodward's chlorophyll had no such functionality, but his mastery of chlorophyll a is one of his greatest achievements. Woodward described the molecule as "a chemical fairyland," and its synthesis—including that symbolic use of light in the final stages—wrung atomic insight from a molecule that conquered the world.

Technology has enabled synthetic chemistry to mature, even though chemical tinkering in some ways has not changed since the days of alchemy. When Koji Nakanishi wrote that structural studies on the ginkgolides were "the last classical and romantic study," he was referring to powerful tools developed in the middle of the twentieth century to probe the world of atoms with great precision. Discovering the ginkgolides' structures would have been impossible without them. It's these technologies that allowed chemists to go from simple starter molecules to ginkgolide B. Each involves beaming molecules with specific kinds of energy, detecting their rattle, and drawing on a career of knowledge to determine a chemical structure.

There are four main tools at their disposal. A mass spectrometer reveals the mass of a molecule when material is injected into a vacuum chamber, and picks up a charge. The molecule soars through the machine to a detector, riding a specifically tuned magnetic field. Once that field is determined, scientists backtrack to the size of the molecule able to ride that frequency.

Infrared spectroscopy measures the vibrational energy of molecular groups. Atoms bound together in small groups emit infrared energy at known frequencies (a key feature for understanding climate change). By measuring this energy, a scientist can tell if carbons are bonded or double bonded to oxygen, what that pair of atoms is bonded to, what oxygen and hydrogen might be doing together, or any of the infinite other associations of atoms.

Neither of these tools is enough. From the mass of a molecule, scientists can calculate its formula. But formulas reveal little about a molecule's structure. Ginkgolide B has twenty carbon atoms, twenty-four hydrogen, and ten oxygen, which can produce more structures than is worth contemplating without more information (such as that gleaned through combustion analysis). Infrared studies reveal what kinds of molecular groups are involved, but not how many or where they are.

The third tool is a nuclear magnetic resonance imager (NMR), which identifies the location within a molecule of nuclei that have an unequal number of protons and neutrons, such as hydrogen (a single proton) or carbon-13 (which has an extra neutron). These unbalanced nuclei become subatomic magnets. An NMR creates a magnetic field around each of these nuclei, and the field felt by each atom is altered depending on the arrangement of atoms—the chemical structure—in the molecule. By measuring the difference between an atom's behavior in a magnetic field and in its absence, chemists can determine the specific environment of the atoms, and thereby assemble a complete picture of the molecule's structure.

X-ray crystallography is the fourth tool, and delivers the most precise measurement. In cases where a material crystallizes, scientists can shoot x-rays at it, which diffract when they hit the electron clouds. The shortwave radiation penetrates structures at the atomic and molecular scale. Watson and Crick deciphered DNA's structure in part by studying the x-rays taken of DNA crystals by Rosalind Franklin. The limitations on this tool come from the materials, which aren't always so photogenic. Crystals can be prohibitively difficult to make for many substances.

The most important supertool isn't a machine in the traditional sense. It doesn't output data on graph paper. All the human brain needs is a writing tablet and pencil. Synthetic chemistry's greatest victories over nature's most forbidding molecules occur in the heads of chemists, where they mash thousands of possible reactions through a sieve, seeking the most elegant path from complexity, backward to simplicity.

With pencils and paper, the synthetic chemists plot their "attacks" on molecules. The revolution at Harvard and elsewhere in the 1950s and early '60s was a revolution in thought, called retrosynthetic analysis. For the previous one hundred years chemists fiddled in the dark, choosing a starting material and trying to perform reactions on it that each took them incrementally closer to the target molecule. Retrosynthetic analysis has chemists mapping their work on paper, backtracking a reaction series that would have a suitably simple starter molecule. Retrosynthetic analysis tipped chemistry from art toward science.

In 1967, ginkgolide B became recognized as an attractive target for synthetic attack, partly because the *Ginkgo* tree itself has such a rough time making it. Nobody knows why *Ginkgo* should go through so much trouble to make these molecules in its roots. Another reason for trying to synthesize ginkgolide B was that it was revealed to work as an anti-PAF in the mid-1980s.

It has been synthesized twice. The first time took E. J. Corey's lab three years, after he had doodled out part of an answer fifteen years prior. The second ginkgolide B synthesis took Michael Crimmins of the University of North Carolina–Chapel Hill more than a decade. The more complex a molecule is, the more paths there are to it. And every synthesis bears the stamp of its chemist.

Ginkgolide B itself doesn't translate well into a flat chemical diagram because its beauty plays in three dimensions, a lovely transposition of rings along six different planes. With its functional groups attached, ginkgolide B is a natural complexity. No human mind could have designed and engineered it. This molecule is a product of biochemical nonlinearity. Since it wasn't engineered, it wasn't meant to be reverse engineered. It just happened once. The genome of the *Ginkgo* survived because some animal liked the nuts or the tree handled the climate well or certain insects couldn't kill it. The metabolic pathway to build ginkgolide B traveled down the evolutionary bend with the other qualities that let this species, the last remaining species, survive on the edge long enough for humans to discover its beauty and utility.

The skeleton of the ginkgolide B structure is its six tumbling pentagons. Of its twenty carbon atoms, eleven sit at so-called chiral centers. That means eleven carbons—aligned contiguously throughout the molecule—can attach to the structure in either of two mirrorlike configurations; all told, that creates a pool of 2^{11} possible combinations—2,047 of which are not found in ginkgolide B. Four of the carbon atoms are fully substituted, or bonded to four nonhydrogen atoms. Two carbon atoms each belong to five of the six rings, perhaps the strongest indicator of how tightly all those atoms are jammed together. Oxygen is an explosively reactive element, yet the synthetic chemist must graft ten oxygen atoms. Half of the carbons bond to oxygen and must do so without destroying other parts of the molecule. The final trap is a tertiary butyl group—a carbon atom that bonds to three methanelike groups. *Ginkgo's* tertiary butyl is the only such structure of its kind known to appear in nature. Through all of this, the designers must make sure atoms' electrochemical proclivities react in such a way not to throw the structure out of whack.

It's a thorny little molecule.

Molecules gain complexity exponentially, given their size, atomic composition, functional groups, ring arrangement, chiral points, chemical reactivity, and stability.[19] Ginkgolide B packs more puzzles into a 420 Da molecule than any known competitor in that weight class. Every criterion for complexity is there: atom types, molecular sidecars, ring arrangements, chirality, and reactivity.

Every September, a company called Garnay Incorporated harvests leaves from the largest *Ginkgo biloba* forest to grow on Earth for 90 million years. Leaves are shredded from stalk, dried, baled, and shipped to Europe, where a French-German joint venture refines the green fans into grains, jams them into capsules, and caps them in amber bottles. The forest, really more than 1,100 acres of shrubs no taller than 3 meters, will regrow in the spring. Sales of ginkgo biloba dietary supplements top $1 billion a year.[20]

Ginkgolide B is the world's best-selling herbiceutical. It sells or is prescribed under different names around the world: Tanamin in South Korea,

Tanakan in France, Rökan in Germany, and BioGinkgo in the United States. They all contain the extract abbreviated as EGb 761. Ginkgo is used to treat faulty memory, spotty concentration, impotence, depression, dizziness, tinnitus, and headaches, often associated with the onset of Alzheimer's or other forms of dementia. A study on the rat brain in 2006 suggests that ginkgolide B might help memory by freeing the passage of glutamate, a neurotransmitter that falls into disrepair in Alzheimer's patients.[21] Research into the biochemistry of its reported effect on early-stage Alzheimer's is still young.

After *Ginkgo biloba* died out more than 7 million years ago in North America, they didn't return until the end of the eighteenth century. Its traditional medical uses remained and to a large extent remain an Eastern tradition. The Food and Drug Administration does not regulate ginkgo, though the FDA and other agencies are studying ways to regulate it.[22] Its potential uses are still debated in scientific journals—and manufacturers cannot guarantee a consistent amount of ginkgo's key ingredients in every pill.

Ginkgolide B is ginkgolide B, whether it came from an uprooted tree, the leaves of a bush clipped every year by farm equipment, or a chemistry laboratory. Corey's or Crimmins's syntheses demand too much time and labor to be of commercial use. But their inventions of unnatural pathways toward a natural product offer a scientific irony. Scientists trained their minds on ginkgolide B and literally know it inside and out—but not how it works precisely inside ailing minds.

9 | Faster Than a Speeding Bullet

Antiballistic Carbon

> "Consider," Panurge went on, "how Nature inspired him to
> arm himself, and what part of the body he first began to
> armour. It was, as God's my life, his ballocks"
>
> —François Rabelais

Duck! That's more or less what evolution grants us as defense from incoming projectiles. We can't grow our own shields. We have skeletons, unhelpfully, on the inside. Reflexes are our best natural defense against sticks and stones and sucker punches, when running away isn't an option.

These limitations became pronounced more than 1,000 years ago when humans started spitting flame-thrown projectiles at each other. Foggy historical records put the invention of black powder in China in the ninth century. Chinese "fire lances" and "hand cannons" matured in fits and starts over the centuries. Black powder's first documented use in rocketlike weapons waited until the thirteenth century.[1]

To fire a weapon is an act of violence. Cannon fire. Pistols fire. An explosion in an open-ended tube, restrained by steel, bestows unnatural speed on small-caliber-scale missiles. The bullet's flight ends in damage every time. The bullet is damaged; the impact site is damaged. Bullet makers and shooters want strong bullets that maximize damage to targets. Weapons grow more dangerous every year. Hollow-tipped projectiles peel open on impact, their jagged edges slicing the victim with multiple tiny knives. Explosive tips detonate on impact. "Cop killer" bullets contain hardened steel plugs that rip through soft body armor and the person wearing it. Potential human targets want shields that inflict maximum damage on the bullet.[2] It's an evolutionary arms race. Makers of ballistic protection can't invent fast enough. Arms makers in China or Russia are all too willing to make ammunition illegal in the United States. Soldiers serving abroad wear ceramic plates in front of soft armor, to deform high-power bullets from AK-47s before stopping them. Soft body armor itself offers protection from only small-arms fire.

How can a fabric vest stop a speeding bullet? Two leading types of bullet-resistant materials draw their strength from crystalline carbon, one reminiscent of diamond, the other of graphite. These materials are powerful illustrations of how carbon works at the molecular level.

Chemistry drives guns, just as it drives automobiles. A thirteenth-century Franciscan friar advised Pope Clement IV that mathematics and natural philosophy—the grab bag phrase for premodern science into the nineteenth century—were the best hope for strengthening the church institutionally and its potential to defeat the Antichrist. Legend holds that Roger Bacon also dispatched to Clement IV the West's first recipe for gunpowder, which called for "salis petrae, *luro vopo vir can ultri* et sulfuris." The strange phrase is a Latin anagram for the recipe's proportions and the third ingredient: black carbon.[3]

The anagram's authenticity and its attribution to Bacon are disputed. Regardless, the friar in 1267 did set down on paper Europe's first description of the "devil's distillate." Bacon called gunpowder "a child's toy of sound and fire made in various parts of the world with powder of saltpeter, sulfur and charcoal of hazelwood . . . By the flash and combustion of fires and by the horror of the sounds, wonders can be wrought and at any distance that we wish—so that a man can hardly protect himself or endure it."[4]

Before the four-stroke engine came along, inventors even tinkered with black powder engines. Christiaan Huygens was one of several seventeenth-century inventors who tried to ignite gunpowder in a sturdy, small space, where its combustion might be captured and transformed into mechanical work.[5] But Huygens failed to find a way to clear the waste gas from his engine, or pump in the next round of fuel.[6]

Black powder is usually a mixture of 75 percent saltpeter, 15 percent carbon, and 10 percent sulfur.[7] The oxygen supply, or oxidizer, is chemically locked within saltpeter. A spark frees the three O's from saltpeter, KNO_3, or potassium nitrate. The sulfur and carbon are sitting ducks. The heat of ignition raises the temperature beyond 261°C (about 500°F), the threshold at which sulfur burns.[8] The sulfur combustion heats the mixture, further, freeing more oxygen from the saltpeter, raising the temperature, and igniting the charcoal. The carbon reactions release even more heat, and so it goes. The gases that formed take up much more room than the solid powder, and press outward at 15,000 pounds per square inch. The fastest way out is through the barrel, behind the bullet.

Black powder proved energetically wasteful because it turns into gas and soot, in similar amounts. The production of all that soot consumes energy that might otherwise be spent on the bullet.[9] Besides, heavy black dust builds up and clogs the gun. Every shot throws up a smokescreen, which can be a plus or a minus. In a gunfight, the shooter can hide behind the dust. But if he's a sniper, the powder blows his cover.

Smokeless gunpowder, also known as guncotton, came about the way so many discoveries do—by accident. Guncotton incorporates the nitrate into a carbohydrate, dispensing with the sulfur. It harnesses that extra energy into the force of the explosion. Guncotton's invention is legendary, both in the sense that it's a great story and because it may not be true.[10] In 1846, Christian Schoenbein, a Swiss-German chemist, spilled nitric acid in the kitchen, despite his wife's instructions not to engage in any chemistry beyond the culinary. He did what any responsible husband would do. He wiped it up with his wife's apron. He slung the garment over the stove to dry, which it did—and very quickly. Heat vaporized the apron. In subsequent decades, other chemists would improve on Schoenbein's accident. Guncotton came to market in the 1880s.

Still ninety more years had to come and go before humankind could, in Roger Bacon's words, "protect himself or endure" gunshots. Inventors took stabs at body armor through the ages. Whoever first slung bearskin over his chest to lessen the impact of rocks thrown at him was on to something, at least until the advent of flint spearheads. Smelted plate mail no doubt looked smashing in medieval military ceremonies, but on the battlefield it may have been an invitation for trouble. Even if an armor-clad warrior, immobile, didn't fall off his horse and drown in muck, he was a favorite target for the enemy. Legend has it that everybody wanted to kill the rich guy in the flashy armor. Poor guys fared badly, too. Woe were Civil War soldiers whose only protection from lead projectiles was skin and shoddy, the recycled wool fabric of their uniforms. Guns drove military planners out of the armor business altogether by the eighteenth century, because no one could invent a suitable defense from bullets. Rifles before this time were big and heavy. When enemies stopped wearing makeshift armor, the guns got smaller.

The solution to stopping bullets harkens to the advent of biological shields. Recall that long before the Cambrian period, algae developed genes that prevent calcification. Millions of years later, when controlled calcification became desirable in the face of Cambrian predators, species after species reached into their genetic tool kits and rewired the old pathways for new benefits. "Exaptation"—the use of a trait for something other than its original function[11]—makes a fine metaphor for the invention and production of bullet-resistant materials. They came about in a search for extraordinary fibers that might strengthen radial tires without using steel. Only after the discovery did the idea of a superstrong fiber independently find its way into the ballistics arena.

E. I. du Pont Nemours and Co. was founded on the bank of Brandywine Creek, near Wilmington, Delaware, in 1802. The company's founder, Irénée du Pont, wanted to build a company that provided something America sorely needed: quality gunpowder. Du Pont was a chemist and trained in France with Antoine Lavoisier, the father of modern chemistry. Du Pont's insight came during a hunting expedition, when he found the American black powder in his gun proved inferior to the European products he was used to. The year after its founding, DuPont dispatched its first commercial batch of gunpowder across the Atlantic, to fill French guns aimed at British and German soldiers. Irénée du Pont boasted to President Thomas Jefferson that his gunpowder would shoot bullets 20 percent farther than English or Dutch bullets. The U.S. military became a customer around 1805, when an expedition put down revolts in the Barbary states of North Africa. DuPont sold 750,000 pounds of gunpowder to the United States between 1811 and 1813, during the war with the British. A century later at the height of its power, DuPont was sued under antitrust laws. Within a decade, Pierre du Pont repositioned his family's company as a player in the emerging chemical and materials industry. In 1903, the company opened its Experimental Station, with about a $75,000 budget, to help it transition out of the powder business.[12]

The chemical industry rose from virtually nothing before World War I, to a fledgling field shortly afterward. The war is less familiarly called the "chemists' war." On the killing fields of Europe, chemical warfare debuted, as did synthetic chemistry. Germany might have surrendered in 1915—instead of November 1918—if its *über*chemist Fritz Haber hadn't developed a way to make nitrates for gunpowder in a factory.[13] To the victors belonged the spoils. After the war the United States auctioned off the rights to German chemical patents and intellectual property that Allied armies took possession of, a boon to industry here.

Shortly after the war's end, a German chemist named Hermann Staudinger made important theoretical advances that propelled industry forward. He described the composition of polymers, which are large molecules of repeating units. Inventors and scientists suddenly had a world of new toys to play with, molecular ones, with which they began to build new pharmaceuticals, plastics, and other materials. Facing pressure to invent novel synthetic chemicals from German and British chemical giants, and from General Electric at home, DuPont expanded its research budget by a factor of fifteen in 1927, priming its Experimental Station in Wilmington to compete.[14]

A molecule-making competition ignited among academics and their upstart industrial counterparts. Plastics chemistry was still an experimental science, with groundbreaking theoretical research laid down after the war.

A polymer is a molecular chain of many ("poly") units ("mers"). An industrial polymer might be thought of as a railroad train. Each car, one the same as the next, hitches to its neighbors. In a polymer train the individual car molecules are called monomers. Proteins are natural polymers, made from life's twenty amino acids, as are materials like cellulose, a polymer of glucose that makes up paper.[15] Polymer isn't a very specific word, which is why so many natural and industrial materials qualify.

By the mid-1920s chemists were building polyesters—the stuff of cotton-blend shirts—with a molecular weight of more than 4,000 Da. (A dalton is one twelfth the mass of a carbon atom, or about a proton.) DuPont plucked Wallace Carothers, a brooding talent from Harvard, specifically to help push the molecular weights of polymers higher.[16] Carothers was sure he could beat the going record, which he did within a couple of years, finishing a polyester with a mass of about 5,000 Da. Practical breakthroughs came in the early 1930s, and not without corporate pressure. Carothers's lab made an advance that led to the development of neoprene, the synthetic rubber today used in wet suits, laptop covers, and all sorts of industrial applications.

In 1931, Carothers created a predecessor to nylon, twice the size of the then-longest polymer. Its peculiar properties lit public attention. The material snapped when chemists applied abrupt force to it. But stretching it produced a thin transparent fiber. Once stretched seven times longer than its original length, the gunk became elastic. Stretching caused the individual molecules to orient in the same direction, and reconfigure into a crystal.

Newspapers extolled the discovery of a polymer that could serve as synthetic silk—if the lab could lower the cost of production. Three years would pass before Carothers figured out a way to make the substance more cost effective, and with the desired properties. He melted the chemical constituents together and pressed them through a glorified showerhead, called a spinneret, where a piston forces the material through pores one one-thousandth inch wide. The material was squeezed out as fiber and landed in a cool bath, ready for processing.[17]

Nylon wouldn't appear to have anything in common with a hamburger, chicken, or tofu. For practical purposes, it doesn't. But like the proteins that make up our bodies or the foods we eat, nylon is a polyamide, or a string of amide molecules. Ribosomes in living cells cobble together amino acids—sequences of the twenty that life prefers—forming a polyamide chain. Electrochemical forces fold the polyamide into a specific shape, the protein, which runs off to perform some function in the cell. Nylon is a polymer of molecules that are practically amino acids. Nylon's basic molecule, or monomer, is itself more complicated than the twenty natural amino acids, but the polymer itself is far simpler than the simplest protein.

Nylon found its way into toothbrush bristles and women's hosiery by the end of the decade, and parachutes and tire reinforcements during World War II. The postwar period saw the chemical industry's greatest time of creativity and prosperity. A generation after Wallace Carothers invention of nylon, scientists at the Experimental Station hoped to find something even more remarkable, a fiber with unnatural properties reminiscent of Superman: it might stop bullets.

Out-of-the-blue accidents populate the history of chemistry, but Kevlar is not one of them. DuPont was looking for something special. It was still a surprise to finally find it, but scientists groped with their eyes open. Fears of future gasoline shortages inspired researchers to look for a better, more fuel-efficient tire strengthener than steel. If it also worked in other applications, such as ballistic fibers, then all the better. DuPont had already scored a success with Lycra spandex and Nomex, the self-extinguishing fiber the company first marketed in 1967, a decade after its invention.[18] Nomex is still woven into firefighters' garb, race car drivers' uniforms, and Hollywood stunt teams' clothes. Cotton emits combustible gases when ignited, fueling the fire. Nomex cuts off oxygen to the fiber, extinguishing the fire as soon as the flame is removed. Vlodek Gabara, senior scientist at DuPont's Richmond plant, has business cards made of Nomex. It burns as long as it is directly engulfed in flame. Out of the fire, the burning stops.

Stephanie Kwolek, grew up in a small town near Pittsburgh and graduated from college assuming she'd go to medical school. A job offer from the Experimental Station turned out to be a better fit. She started at the company in 1946, and by the early 1960s she and her colleagues were engaged in the search for a successor to Nomex that might have extraordinary strength.

Kwolek came across a promising-looking substance, called p-aminobenzoic acid, a six-carbon ring with amine and carboxylic acid side groups. (Today it's a common ingredient in sunscreens.)[19] The substance wouldn't melt, as nylon or many others do. So she dissolved it, reluctantly, in sulfuric acid, which the lab prohibited researchers from using. Sulfuric acid is highly toxic, expensive, and tough to dispose of safely. (Today, it is produced in greater quantity than any other industrial chemical, mostly for its importance in making fertilizer.) Nylon melted into a transparent liquid the consistency of molasses. The p-aminobenzoic acid solution looked cloudy and had the density of water. Her colleague in charge of the spinning machine refused her early requests to run the substance through the machine, fearing the cloudy material would clog the holes.

Kwolek ran the substance through a glass funnel, and the material moved through cleanly. Though she didn't know it at the time the "opaque and pearlescent" solution contained the world's first synthetic liquid-crystalline polymer.

She pressed for several days and eventually got her way. "He felt sorry for me," she said. The liquid spun right through. She tested the tenacity, or breaking strength, of the resulting fibers. The results were so startling she immediately retested it three times,[20] hewing closely to Isaac Asimov's observation that scientists don't shout "Eureka!" upon making a big discovery. They mutter, "That's funny . . ." The assumption of error commonly greets scientific breakthrough.

DuPont had found its superfiber. But no one discovers anything in a vacuum, and no one scales an industrial polymer to commercial levels on her own. Without colleagues and a blue-chip industrial company backing it, DuPont's superfiber would never have succeeded. Like the initial attempt at nylon, Kwolek's discovery was too expensive to scale up. The lab went to work understanding the physical chemistry of the molecule, looking for clues, and a way to make a cheaper polymer that didn't sacrifice strength or rigidity. The most promising chemical, abbreviated as PPD-T, had the stiffness of Kwolek's original molecule but called for less expensive ingredients.

PPD-T starts out a golden yellow powder and ends up as Kevlar, a golden yellow fiber wound onto 8.5-inch-tall bobbins. The powder is dissolved in sulfuric acid and forced through the spinnerets. The molecules are too numerous and rigid to sustain random ordering. The more PPD-T is added to the solution, the closer the molecules get to each other and—they're so rigid—the harder it is for them to maintain random orientations. They are forced into tight alignment and end up running parallel to each other, in a liquid crystal. This crystallization of liquid was the fundamental scientific breakthrough behind the invention of Kevlar, and other modern conveniences, including LCD (liquid crystal display) televisions and monitors.[21]

Aramid fibers go some way to diffuse a seeming paradox about how order can emerge without violating the second law of thermodynamics, which states that entropy, or disorder, tends to increase over time. Like lightning or hurricanes, or the emergent primordial metabolism posited by Harold Morowitz, molecules under the pressure of a spinneret can become ordered as a way to disperse energy they would otherwise have to spend bouncing around the solution. Those rigid-rod molecules shouldn't have any particular need to crystallize. Energy, in this case via mechanical pressure, directs those molecules through the spinneret. As their concentration builds, crystalline alignment becomes a more favorable way to dissipate energy. PPD-T molecules share their energy in chemical bonds because it's an easier way to shed energy than haphazardly bumping into each other. Sometimes order is the most efficient way to move energy along. Aramids form because chemical bonding becomes an easier way to cope with energy than random molecular bumper cars.

Finding the right chemical still didn't solve the problem. Manufacturing was still too impractical. Engineers balked at the requirements for designing, building, and operating a plant that would pump a river of 100 percent sulfuric acid through its pipes every year.[22]

The last shoe to drop before the engineers signed on came after Herbert Blades, a top scientist at the Experimental Station, figured it out. The spinneret enforces a structure on disoriented polymer chains, a little bit like the way a box of spaghetti encourages dry strands to remain side by side, facing the same direction. The higher the concentration of the polymer in the solution, the tighter they are forced to jam together and the more packed the spaghetti box. But as the fibers exited the spinneret they inevitably lost some of that structure as they hit the cold-water bath below. Blades discovered that leaving an air gap between the spinneret and water bath gave polymers time to reorient themselves back into the unidirectionally oriented superpolymer. With the air-gap discovery the economics looked more favorable. A task force of scientists and engineers went to work on a start-up plant in Richmond, Virginia, which used to be the site of a thriving textile industry. Soon it would become the hub of a thriving supertextile industry. Two years after Blades's air-gap discovery, the plant churned out a million pounds of Kevlar a year, an unprecedented start-up given the sophistication of the material, fabrication, and the quantity of toxic waste neutralized. Five years later, output reached 15 million pounds.[23] Kwolek monitored the scale-up, but otherwise continued her research at the Experimental Station. The Kevlar name came later, when creative types sat around looking for a masculine-sounding name that didn't mean anything in any language. Profitability came still later, in 1988, after DuPont had funneled more than $700 million through spinnerets ($1.35 billion adjusted for inflation).[24]

Two leading antiballistic fibers owe their strength to variations of carbon's two kinds of carbon crystals, diamond and graphite. Kevlar and Spectra are DuPont's and Honeywell's trade names for the generic chemicals called aramid and ultra-high molecular weight polyethylene.

Diamond's hardness comes from the shortness of the carbon-carbon bonds and their three-dimensional orientation. Electrons "like" to move as far away from each other as possible. Carbon's electrons prefer to be arranged in a tetrahedral pattern, at angles 109.5 degrees from each other, approximately the observed angles in a diamond lattice. The atoms are stuck in place, welded to four other carbon atoms in precisely the same situation.

The diamond lattice is an obstacle course for white light. Imagine trying to lead a phalanx of sport utility vehicles through a mountain forest. Every driver makes it through, but each slows to his own speed. The fearsome symmetry of the phalanx is shattered. That's what happens to light in diamond. Light is a stream of particles called photons. White light breaks down in a prism to its component colors that are remembered by school children as Roy G. Biv (red, orange, yellow, green, blue, indigo, violet). Every color in the spectrum has its own tread width, or wavelength, measured in hundreds of nanometers. Just as people see white light, scientists can "see" every other part of the electromagnetic spectrum.

Roy G. Biv makes up about 4 percent of a spectrum called electromagnetic radiation. Visible light, the band of wavelengths emitted most powerfully by the Sun, sits in the middle of that range. Wavelengths longer than visible light include radio waves, microwaves, and infrared light. Ultraviolet light, x-rays, and gamma rays carry shorter, more powerful radiation. The heat trapped by the atmosphere, and that our bodies emit, is long-wave radiation.

All of these forms of radiation, including white light, travel at 186,000 mps in a vacuum. Outside a vacuum, light hits speed bumps. In a diamond, light has all the trouble a cheetah would have running through knee-deep molasses. The speed of light plummets to 80,000 mps when photons cross the threshold between air and diamond.[25] Carbon atoms are packed so tightly that color wavelengths trip over different atoms and fall out of formation. The result is that photons bounce around the many facets looking for a way out. We see the colors of the spectrum skipping one into another, bouncing off the walls, seeking an exit. Each color finds its own egress, whenever it can.

Scientific speculation about diamonds' composition dates at least to Sir Isaac Newton. He proposed that diamond must be related to substances that burn. Perhaps diamond is an "unctuous body coagulated," he wrote, meaning a congealed oil.[26]

In 1772, Antoine Lavoisier sealed about 150 mg of diamond in a container. A carefully placed magnifying glass channeled the Sun's heat onto the glass. Analysis revealed the container afterword to contain only air— but more air than Lavoisier started out with. The vaporized diamond mated off with oxygen in the air to become carbon dioxide, the same gas we exhale and power plants emit. Lavoisier did not call it that, because he would not coin the word oxygen until 1774. In a follow-up experiment, he placed 150 mg of charcoal in a box. It too vaporized into the same gas, in the same quantity. Diamond and charcoal were made out of the same substance, he concluded. Fearing his colleagues might find such a discovery ridiculous, which it is, he didn't broadcast his results. British scientists extended this work half a century later.

Ultra-high molecular weight polyethylene (PE) shares properties that make diamond so strong. In the late 1970s and early 1980s, Allied Signal scientists looked at their portfolio of industrial fibers and itched for the next big thing. Nylon and polyester fibers could be manufactured into anything from T-shirts to industrial-strength ropes. But the nylon and polyester molecules are complicated, which leads to more defects in production. Nylon and polyester are also short polymers, which makes them more likely to resist crystallization and entangle. Their tenacity, or breaking strength, is low. Ripping a T-shirt is easy. Pulling steel apart is easier than an equal weight of ultra-high molecular weight polyethylene—ten times easier.

Like their DuPont counterparts several years earlier, the scientists were looking for a simpler polymer that would crystallize better, one that could fill immediate market needs for rope, industrial applications, and in emerging markets, such as ballistics. They found it in polyethylene, a common plastic that exhibits extraordinary properties in ultra-high molecular weight sizes. The strength comes from the simplicity. Ethylene is a two-carbon molecule with four attendant hydrogen atoms. Polyethylene is a straight chain of carbon atoms, each of which is flanked by bonds to hydrogen atoms. Long-chain polymers got a boost in the middle of the last century when chemists learned how to build extremely large ones by using metal catalysts, rather than heat and pressure. The discovery of "clamshell" catalysts—thus nicknamed for the regularity with which they "open" and then "close" on the next polymer bond—lowered the costs of making polymers and proved to be the starter's pistol for the age of plastics. Lori Wagner, Honeywell's top Spectra scientist, likens ultra-high molecular weight polyethylene to a diamond, stretching out in one direction. Its strength is perfectly analogous to diamond, carbon-carbon bonds running up and down, right and left across the vest. Its high energy-absorbing potential comes from layering fibers perpendicularly.[27]

Allied Signal scientists invented a process for making fibers of great length. The polyethylene is placed in a gel solvent. That loosens up the entangled molecules enough that they can be pushed through a spinneret, which imposes order on the polymer and orients its component molecules up and down the length of the fiber.

Polyethylene vests are not woven. Initially the company wove vests from the polyethylene, but the weave inhibited the strength of the crystals oriented along the length of the fiber. The fibers are laid out in sheets. The sheets are fixed one atop another with a resin, each layer's fibers oriented perpendicular to the layer below. Laterally, the individual polymer strands are bonded together in the fiber through van der Waals forces, quantum-world effects stronger than any glue. This gives ultra-high molecular weight polyethylene vests their viscoelasticity, a fancy word for a simple concept. The harder something hits a polyethylene vest, the harder it hits back.

Graphite is also a carbon crystal, though you can't tell this commonality to diamond by looking at them. Carbon atoms in diamond are arranged in tetrahedral fashion, each carbon sits at the vertex of 109.5-degree bond angles made with four others. Graphite is black and shiny. It's softness makes it great as pencil "lead" and as an industrial lubricant. These uses

August Kekulé in 1862.

depend on a wholly different arrangement of carbon atoms in the crystal. Graphite is best pictured as layer upon layer of stacked carbon sheets. These sheets, called graphene, are composed of six-atom carbon rings, linked together in what is usually depicted as interlocking hexagons. Graphene sheets are extremely strong, in fact the most stable arrangement of carbon atoms. But they are stacked together only loosely, which helps explain graphite's smudginess.

A good place to start explaining why graphite is so stable is by looking at one of those six-carbon rings, which is the central structure of benzene, one if the most ubiquitous structures in nature and most storied molecules in science.

In 1866, August Kekulé solved the structure of benzene, C_6H_6, a ring of carbon with hydrogen atoms radiating out. Kekulé delivered a lecture at a twenty-fifth anniversary celebration of his greatest discovery, in which he told one of the more dramatic and enduring stories in the history of science. Though he hadn't mentioned it to anyone over the previous twenty-five years, Kekulé said that the structure of benzene came to him in a dream. He saw a snake winding around and biting its own tail, an ancient, mythical image, known as Ouroboros, a symbol of the eternally consuming and creating nature of life. The image led him to understand that the six carbon atoms in the benzene are bonded in a highly stable chemical ring. Ouroboros is much more than a symbol for life, but science itself, which progresses by destroying and re-creating itself—perhaps one reason the benzene ring is such an iconic chemical symbol.

Whether or not Kekulé really had his dream is a dispute so impassioned that it still caused polemics and even lawsuits a century later and half a world away in the United States. In a *Lancet* review of a book of anti-Kekulé essays,[28] the reviewer concludes of the tome and the bitter polemicists, "Nobody considers the possibility that he had a sense of humor . . . Let us give Kekulé credit, then, for inventing such a brilliantly memorable ripping yarn."[29]

Within a generation after the discovery of its structure in 1866, chemists already determined that these rings, or permutations of them, roll through more than 75 percent of organic molecules, including the dyes and pharmaceuticals that dominated industry. Most of the molecules discussed in this book are built on polycyclic hydrocarbons— space soot, the hopanes resembling cyanobacteria molecules, ginkgolide B's cascade of six rings, adrenaline, tetrahydrocannabinol, lignin, aramid. Carbon-based rings can have oxygen or nitrogen atoms plugged in, such as in DNA's nucleotide bases. Carbon's ring-making ability helps make the DNA nitrogenous bases the flat little pieces of cardboard that they are.

Benzene and graphite's molecular strength comes from an unusual arrangement of carbon atoms, which Kevlar takes advantage of, too. As we've seen, carbon's architectural magic emerges from its ability to make multiple bonds with other atoms. Two atoms are typically connected to each other by sharing two electrons, one from each atom, in a single bond, such as the $C-C$ bonds in a diamond or in the backbone of saturated fats. Sometimes atoms share more than one electron with each other, including the $C=C$ bonds in unsaturated fats or the $C=O$, $C=N$ double bonds that serve as reaction points in life's reactive molecules. When atoms each share two electrons with one another, they are doubly bonded.

The same goes for three electrons, the strongest bonds, evidenced in those long carbon chains that appear in the dense interstellar clouds, the cyanopolyynes—HC_5N, HC_7N, etc. Their carbon atoms alternate triple and single bonds in a linear arrangement.

The secret of benzene and graphite—and aramids—is that their carbon makes bonds that are neither single, double, nor triple. Benzene rings and graphite draw their strength from special molecular structures called *pi* bonds. Benzene rings might be thought of as molecular bagel sandwiches. Every carbon bonds to two others, all the way around the hexagon. Those bonds occupy two of each atom's four unpaired electrons. The carbon atoms lie in the plane of the cream cheese. Each carbon is also bonded to a hydrogen, which radiates out from the ring, in the same plane. That takes care of three carbon electrons, and it explains the six-carbon, six-hydrogen molecular plane of the benzene ring. But carbon has four electrons to bond with.

The fourth electrons inhabit the bagel-shaped space above and below the ring. Typically, electrons would stick close by their home nucleus. But in benzene, the electrons "delocalize." Each carbon atom has responsibility for holding the electrons of adjacent *pi* orbitals together. That's what makes benzene rings such powerful structural materials. The rings don't stretch. The molecules vibrate like others, but without jeopardizing that ring. Benzene molecules have "breathing mode," in which the carbon-carbon bonds expand and contract, presumably like the cross section of an expanding rib cage. Single hydrogen atoms bonded to each carbon sway in the opposite direction of the twist, the way a TV shampoo model's hair sways in the opposite direction of her head.

Graphite can be thought of as sheets of benzene rings. Individual sheets are called graphene. They stack one atop the other in a staggered manner. In graphite, the electrons in *pi* orbitals delocalize among all the carbon atoms, making those sheets even stronger than one molecule of benzene. Yet the sheets themselves are bonded together only loosely.

The strength of aramid molecules comes in part from these crystal-lized aromatic rings. In Kevlar, carbon rings link one to another in a long polyamide chain, the rings locked one to another by nitrogen and carbon-oxygen groups in powerful covalent bonds. The three-dimensionality of the structure reinforces the core molecule's strength. Polymers stack, sort of like dinner plates, carbon ring atop carbon ring. That gives aramids their strength in a second direction. Finally, the polymers join laterally through hydrogen-hydrogen bonds. In a fiber, sheets of laterally bonded polymers radiate around a central point; a fiber's cross section would look something like a bicycle wheel, the spokes representing fiber sheets.[30]

In May 1972, Alabama governor George Wallace, the notorious opponent of desegregation, was shot five times by a fame-seeking gunman. Months later, Senator John Stennis of Mississippi was shot and nearly killed during a mugging. In July 1973, the air force attaché in the Israeli Embassy to the United States was shot and killed near his house. These high-profile events followed a decade during which police fatalities steadily rose and political assassinations shrouded the nation in mourning. As director of the National Institute of Justice (NIJ), Lester Shubin was prepared to do something about it, and an exciting opportunity had just presented itself.

Shubin knew his way around a firearm. As a combat engineer in southern Europe during World War II he "picked up mines, built bridges, dug trenches, laid barbed wire, fixed roads and got shot at . . . A lot!"[31] His job at the NIJ was to oversee and standardize technologies that help law-enforcement officers do their jobs. He had been administering a program at the Aberdeen Proving Grounds in Maryland that trained dogs for use in bomb sniffing. In 1971—coincidentally the same year DuPont announced it would stop making gunpowder—Nick Montanarelli, an army officer running the dog program for him called with interesting news. The army had been experimenting with a new fiber designed to replace steel-belted tire reinforcements. It might make a good antiballistic fiber, too.

Shubin and Montanarelli wrapped a phone book in Kevlar and took turns firing rounds at it. "The bullets bounced off," Shubin recalled.

The NIJ gave Shubin some money to conduct rigorous tests on the material. The goal of the research was to determine if the material was strong enough to stop bullets shot at police officers. The testing was too important to rely on dummies, cadavers, or other proxies for flesh and blood law enforcement. The researchers wanted live test subjects to see not only if the armor would stop bullets, but to make sure that the impact caused no body trauma, even without penetration. They pressed goats into service.

Goat anatomy is similar enough to human to stand in. Military surgeons became interested in the project and signed on to examine the goats after their experience as Kevlar guinea pigs. The Edgewood team shot the goats with .22s, .38s, .357s, 30-aught-6s. They had to determine the velocities of bullets from these guns, which hadn't been done before. A five-ply vest proved too weak. A .22 long rifle at a 30-degree angle penetrated easily. They added two layers, tightened the weave, and raised the fiber strength. It stopped the .22.

Army surgeons were only too happy to participate in a study as novel as the ballistics testing of Kevlar. All the goats survived that round of anatomical tests, but were put to death the next day so surgeons could check for internal damage. Comparing human and goat anatomies, the surgeons extrapolated how damage to the goats might translate into a human vest wearer. They concluded that a seven-ply Kevlar vest reduced the likelihood of death from a .38 caliber bullet from up to 25.4 percent to 1 to 5 percent. The need for surgery was reduced from 81.5 to 100 percent of the time to between 7 and 10 percent.[32]

The government had a rough time persuading police departments to distribute vests to officers in some parts of the country, where wearing even a long-sleeve shirt is too hot. Asking officers to wear a heavy vest was a bit excessive. Still, Shubin found fifteen departments around the country that took his five thousand vests in 1975. In the first month of service, three lives were saved. After that few needed further convincing. The proliferation of vests had immediate effects. The number of police officers killed in gunfights began to fall annually.

Thirty years later ballistic threats increase quickly and reliably enough to never give armor designers a moment's peace. Today, explosive-tipped bullets blow through protection. Hollow-tipped bullets peel on impact like petals of a steel flower. Ballistic testing has become a highly technical, standardized endeavor.

Twin ballistic testing labs coexist within several miles of each other in Richmond and Petersburg, Virginia. The rooms look similar, about 75 feet long, 20 feet wide, and an office building ceiling tall. The back wall is made of cement two feet thick. Ventilation systems draw toxic lead released into the air on every impact. The ballistics labs use automated firing mechanisms, not actual guns. DuPont's trigger dates to the 1920s, when the company tested its gunpowder with it. Technicians switch out barrels depending on the ammunition they are firing and the vest they are testing. The firers don't look much like guns but that doesn't matter. Energy equals half the mass of a projectile times the square of its velocity. A 0.0156 kg bullet traveling at 436 m/s—about ten times faster than a major league fastball—delivers enough energy into an antiballistic vest, 1,483 J, to knock a smaller police officer off her feet.

At Honeywell's testing facility the technicians dress up the firing mechanism to shoot a .44-caliber bullet—essentially a Dirty Harry size gun—at an aramid vest, a big yellow square several yards away. Honeywell prides itself on its proprietary Spectra polyethylene armor, but manufactures aramid fiber, too. A red laser pinpoints the target on the vest. From a control booth behind the gun, a technician pushes a button. The yellow fabric dances instantaneously to the sound of the gunshot.

The bullet hits the vest's outer layer traveling at 1,425 feet per second. Energy moves through the aramid and PE vests twice as fast as the bullet moves through air. The energy ripples out through the vest at nearly 3,000 feet per second. The interaction between the bullet and target is too complex to easily model, because all the variables involved change with every nanosecond. The bullet rips through the first layer of protection. The second layer absorbs another wave of energy. Speed transforms into heat. The temperature of the bullet rises. Another layer broken, more energy pours through the vest. The bullet sheds mass, and drag increases. The impact with the fiber melts the copper and lead. The surface area of the bullet expands, which slows it down exponentially further. The more the bullet's tip deforms the easier it is for the vest to collect its energy. The deceleration from 1,425 feet per second to zero might take two millionths of a second.

The vest drapes in front of a slab of clay, the stand-in for a human target (goats were released from service in 1977). NIJ regulations demand that the impact push the vest no farther than 44mm into the clay. A postmortem on the clay reveals the vest kicked back 41mm—3mm to spare.

The gun reloaded, a white polyethylene target is hung in front of the clay. Another laser dot is trained and another gunshot. The hollow-tipped bullet shatters into so many pieces, Bob Arnett of Honeywell has to separate the PE layers to find all of them. Then, the moment of truth. Vest makers compete not only against arms makers. They compete against each other. "Please let this be less," he says, in a quiet tone more reminiscent of internal monologue than witty banter. The impact site is just 38mm deep in the clay.

I think: Half a world away, in Iraq, a hydrocarbon war, a 7.62×39 mm bullet from an enemy AK-47 may have just shattered the ceramic plate of a U.S. soldier's body armor, without penetrating it, the plate's fragments neutralized by a cloth vest of crystalline PE.

10 | The Bell Jar

Humans and the Hundredfold Acceleration of the Carbon Cycle

Frequently Asked Question 7.1: *Are the Increases in Atmospheric Carbon Dioxide and Other Greenhouse Gases During the Industrial Era Caused by Human Activities?*

Yes, the increases in atmospheric carbon dioxide (CO_2) and other greenhouse gases during the industrial era are caused by human activities.

—Intergovernmental Panel on Climate Change,
*Climate Change 2007: The Physical
Science Basis*

Civilization has developed, and constructed extensive infrastructure, during a period of unusual climate stability, the Holocene, now almost 12,000 years in duration. That period is about to end.

—James Hansen

Michael Faraday worked as a bookbinder through his teens, reading whatever crossed his path, including the *Encyclopedia Britannica*. Chance tickets to public lectures on science in 1812 became his entrée into the world of professional science. He transcribed the talks at great length, and sent his notes to lecturer Humphry Davy, who eventually hired him into the Royal Institution as a lab assistant.[1]

Chemistry by the early nineteenth century had entered its modern

period, but barely. Tinkerers hadn't even reached the end of the beginning. That came in 1828, when Wöhler made urea without his kidney,
tearing down the barrier between living and nonliving chemistry. The
discoveries that catalyzed the early pharmaceutical industries had to
await the 1860s and thereafter. Not until the early 1930s would modern
physics empower chemists to understand the structure of atoms and
grow more adept at synthesizing complex target molecules from simple
chemicals. Davy and his peers stood at the front end of this pageantry of
chemistry. Davy first linked chemical reactions with electricity. He concluded that some fundamental particles attract each other in "naturally
opposite states," or what we now know are protons and electrons.[2] Davy
was a prominent figure among the many European chemists probing the
apparent electrical nature of chemistry. Since an Italian physicist first
caused severed frog legs to twitch with an electrical current in the 1760s,
European scientists had investigated the connection. In the second decade of the nineteenth century, Davy's great contemporary, the Swede J. J.
Berzelius assembled the system of Latin letter symbols (C, H, O, N, P, S,
etc.) that constitutes chemical shorthand. He also classified the elements
based on how they behaved when confronted with the positive and negative poles of an electrical current. Even today top academic chemists trace
their intellectual lineage to Berzelius.

Still, despite breakthrough insights from Davy, Berzelius, and others,
no one knew electricity and chemistry were the same thing. Michael Faraday made many contributions to science but electrochemistry may be his
most notable achievement. Chemical change, from the ion-molecule reactions of dense interstellar clouds, to the internal combustion engine, occurs because electrons fall in or out of bonds, breaking and making new
molecules as they go. In the 1830s, Faraday showed that a volume of
chemical breaks down in proportion to the amount of electricity channeled into it.[3] He had shown a direct, quantitative connection between
the two previously separate phenomena.

Faraday was also a great communicator, perhaps inspired by the lectures that launched his scientific career. A candle might not seem a portal
into the essence of life, but Faraday made it so in a series of six lectures he
delivered in 1860, called "The Chemical History of a Candle." He expertly
walked his audience through an incandescent encyclopedia of mid-
nineteenth-century physics and chemistry. A candle flame is carbon,

black particles glowing yellow, fed with oxygen and stoked by the parallel combustion of the candle's hydrogen. Faraday concluded the last lecture by illustrating the ubiquity of carbon combustion. The dissolution of carbon and hydrogen, their union with oxygen, characterizes not just candles, Faraday explained, but the symphony of respiration that drives animal life and civilization. He marveled at the quantity of carbon spouted into the air every day. Look, he invited his audience, this is how the world works:

> You will be astonished when I tell you what this curious play of carbon amounts to. A candle will burn some four, five, six, or seven hours. What, then, must be the daily amount of carbon going up into the air in the way of carbonic acid! What a quantity of carbon must go from each of us in respiration! What a wonderful change of carbon must take place under these circumstances of combustion or respiration! A man in twenty-four hours converts as much as seven ounces of carbon into carbonic acid; a milch cow will convert seventy ounces, and a horse seventy-nine ounces, solely by the act of respiration. That is, the horse in twenty-four hours burns seventy-nine ounces of charcoal, or carbon, in his organs of respiration to supply his natural warmth in that time. All the warm-blooded animals get their warmth in this way, by the conversion of carbon, not in a free state, but in a state of combination. And what an extraordinary notion this gives us of the alterations going on in our atmosphere. As much as 5,000,000 pounds, or 548 tons, of carbonic acid is formed by respiration in London alone in twenty-four hours. And where does all this go? Up into the air.[4]

Faraday used an array of candles in his talks, their waxes drawn from beeswax, ox fat, sperm whales, and "paraffine obtained from the bogs of Ireland." Victorian Britain was already lighting the Earth like a candle. Since the late 1700s, Londoners had carried black umbrellas because raindrops collected airborne coal soot and ruined lighter-colored parasols.[5]

If current trends continue unabated, humans will replenish the atmosphere with more carbon than it has had in tens of millions of years. When Faraday spoke, the burning was only starting in earnest. Taking into account electricity use, transportation fuel, and materials, today, the United States emits about 120 pounds of carbon per capita into the air daily, perhaps a hundredfold increase from the Londoner of Faraday's time. Faraday leaves carbon up in the air, unaware that most of it hangs there for a century. About a third of it floats on air for hundreds of years more.[6] The more carbon joins the atmosphere, the longer it stays—and the hotter it gets. Carbon burned from Faraday's candles still inhabits the troposphere.

Scientists have tracked carbon in the atmosphere for two hundred years. By the time Faraday gaped innocently at the "wonderful change of carbon," scientists had concluded that the gases carbon dioxide and water vapor trap heat. Jean Baptiste Joseph Fourier, a Frenchman with a penchant for stifling heat, posited in the 1820s why the Earth retains solar heat rather than radiating it all to space. He performed some calculations and hypothesized that the Earth's atmosphere captures some heat, like a bell jar.[7] The same could be said for a greenhouse, although these metaphors are imprecise. Heat builds up in a bell jar or greenhouse because the air can't circulate very far, not because the glass itself retains the heat.

Carbon and climate studies garnered more and more attention as the nineteenth century went by. Some naturalists were curious about the invisible material that becomes visible once plants absorb it in photosynthesis. Others became caught up in the rigorous debate over the existence and causes of ice ages. They strived to understand (as they still do) what causes the Earth system to dip into cold spells. Scientists reasoned that atmospheric gases fluctuated.

A year before Faraday's 1860 candle lectures, British natural philosopher John Tyndall considered which gases best absorb heat. Ice intrigued Tyndall. He climbed ice- and snow-covered Alps, helping to popularize mountain hiking. He studied the properties of ice, the slow push of glaciers, and the heat retention of some gases. Faraday helped him become a professor at the Royal Institution, where Tyndall would later succeed Faraday as its superintendent.[8]

Tyndall's curiosity led him to study the interactions of atmospheric gases with heat. He invented the first ratio spectrophotometer, with which he observed that water vapor and carbon dioxide absorb heat. The device worked like this. Gases entered a long, thin tube, sealed tightly with salt crystal, an insulator. A heat source at one end of the tube raised the temperature of the gas inside. An apparatus at the other end transformed the heat into an electrical current. This signal, when compared with a control measured from a second heat source showed how much energy the gases inside the tube were absorbing. The heat-trapping action of the molecules larger than O_2 and N_2—namely CO_2 and H_2O—might explain "all the mutations of climate which the researches of geologists reveal," Tyndall wrote.[9]

Scientists did not have precise measurements of the atmospheric gases until the middle of the twentieth century. About 78 percent of the air is molecular nitrogen, N_2; 21 percent is molecular oxygen, O_2; argon makes up 0.9 percent; the rest is mostly water vapor, with carbon dioxide and the other greenhouse gases providing influence greater than their parts-per-million presence might indicate. The effect of water vapor on climate is obvious enough. For example, Caribbean nights are warm because ground-level water vapor retains the day's heat.[10] But we can't feel CO_2 levels the way we can humidity.

Chemistry shot through a growth spurt in the two decades after Tyndall's experiments, as chemists put together a rudimentary understanding of molecular change. The questions of ice ages and climate shifts persisted. Great scientists struggled with them, usually only after they became great scientists. For Svante Arrhenius, recognition took a while. Arrhenius, a Swedish physicist, presented his doctoral thesis about the dissociation of chemical bonds to University of Uppsala professors in 1884. They gave it low marks, but it later pushed electrochemistry into a new era by showing that—as his compatriot Berzelius conjectured two generations earlier—electrical charges drive chemical reactions.

Arrhenius is frequently remembered for a paper published in 1896 that looked at the potential effects of changes in the carbon dioxide content of the air. A drop to 60 percent of then-estimated levels might cause a temperature dip of 4°C or 5°C, he found. A doubling of CO_2 might pump temperatures 5°C or 6°C. He suspected it would take another 3,000 years before coal burning would cause such a significant CO_2 increase. (An estimated 6°C rise at the Permian-Triassic boundary pushed about 95 percent of species into extinction 251 million years ago.)

During Arrhenius's time, measurements of the CO_2 content of the atmosphere were too imprecise to show if the level was climbing. Many scientists tried to determine how much carbon dioxide, then called "carbonic acid," was in the air. Naturalists interested in photosynthesis first raised the question seriously at the end of the eighteenth century.[11] A prominent London chemistry journal published more than 120 articles about the problem over the course of the nineteenth century. Some took what now seem ludicrous approaches to its study. One chemist fastened a 160-gallon chamber to a horse-drawn cart and drove it around town to collect well-mixed air samples. Another sufficed with a 1-ounce bottle of air. However large the sample, chemists had a couple of basic routes to estimate how much CO_2 the air contained. Some filled glassware with air and reactive substances that absorb carbon dioxide, determining the carbon content by measuring either the volume of the remaining air or the change in mass of the reactive substance. Others mixed air in a liquid that dissolves CO_2. The more acidic the liquid, the more CO_2 dissolved.[12]

These and other methods all turned out to be imprecise, into the 1950s. Still, rudimentary estimates, coupled with knowledge of the heat-trapping nature of trace atmospheric gases, allowed rare, forward-thinking scientists to predict global change. G. S. Callendar, an English steam technologist with an interest in atmospheric CO_2, looked through temperature data gathered from two hundred or so stations around the world. Mathematical modeling convinced him that greenhouse gases and climate change were linked. Industrial fossil fuel combustion was increasing the CO_2 content of the atmosphere and therefore raising the temperature. His 1938 article, "The Artificial Production of Carbon Dioxide and Its Influence on Temperature," suggested that industry had belched 150 billion tons of carbon dioxide into the air in the previous half century. "Few of those familiar with the natural heat exchanges of the atmosphere, which go into the making of our climates and weather, would be prepared to admit that the activities of man could have any influence upon phenomena of so vast a scale . . . I hope to show that such influence is not only possible, but is actually occurring at the present time."[13] Callendar's equations led him to think that a doubling of carbon dioxide would cause a 2°C increase, with acute warming at the poles. The scientific establishment dismissed Callendar's work at the time. They would today, too, but for a different reason. His estimate has a 90 percent likelihood of being too low. Today, scientists consider 2°C to be the upper limit of temperature change that might be tolerable. Yet the Intergovernmental Panel on Climate Change (IPCC) has predicted that 2°C is the bottom of the temperature range likely to occur from a doubling of preindustrial CO_2

levels. The IPCC found with equal confidence a doubling might push temperatures up by 4.5°C. It is commonly understood that change is trending toward the higher end of this range.[14]

In the first third of the twentieth century, science was no longer the province of lone inquirers like Faraday and Callendar. Science became more expensive, and demanded more sophisticated equipment and more resources and people. Disciplines became too specialized for one person of reasonable age to leap from one to the next. Professors relied increasingly on their labs to back up and execute their ideas, paid for by institutional or government grants. DuPont's Experimental Station was up and running, along with similar research and development shops at General Electric and other industrial giants. Science became an institutional affair. In the 1950s, Charles David Keeling's graduate work in chemistry at Northwestern University led to an insight that landed him national industry attention: Carbon-carbon double bonds jump to the ends of polyethylene chains when bombarded by high-energy neutrons.

Visiting a friend at the University of Illinois, Keeling pulled a book off the shelf called *Glacial Geology and the Pleistocene Epoch*. The Pleistocene is the stratigraphic period covering 1.806 million to 11,800 years ago, a span of irregular, repeating ice ages. He took a position at the California Institute of Technology, which had just opened a geochemistry Department. The pioneering work of Harold Urey at the University of Chicago led a nationwide academic drive into geochemistry. Under Urey's direction, Stanley Miller conducted his 1953 spark-discharge experiment that launched experimental origin-of-life studies. Harmon Craig, also a student of Urey, standardized measurement of carbon-13 isotopes, choosing belemnite rock from South Carolina's Pee Dee River basin as the standard.

A senior researcher suggested Keeling measure how the carbon dioxide concentration of water balances with atmospheric CO_2. But nobody really knew the latter level, or if there was a stable one. Despite smatterings of work in the late nineteenth century, scientists had yet to study the carbon dioxide content of the atmosphere with any regularity or precision. Conventional wisdom posited that CO_2 made up 300 ppm in a volume of air. Some studies suggested a number as low as 150 ppm, others as high as 350 ppm.

Keeling knew not to trust the CO_2 studies. He needed an instrument to correct them with precision, and searched through old scientific literature to find a way to measure gas pressure. He found instructions for a proto-type in a 1916 journal article. The university glass blower produced a version from sketches, a tool that used the pressure sensitivity of mercury to gauge the contents of an air sample. Air measured atop the geology building showed great variation, but the manometer worked well in local tests. He headed out to Big Sur, a day's drive away, unsure what to expect.

Humans cannot detect changes in atmospheric trace gas composition on their own. Yet it is simple to measure CO_2 and temperature changes with the right instrumentation. This speaks more to a more general principle in science itself. Our bodies perceive reality inadequately. Humans cannot see things smaller than the width of a hair two arms lengths away, cannot walk farther than perhaps twenty miles a day; we blink in about a third of a second, and we live about eighty years. Everything smaller than a hair, longer than twenty miles, occurring more quickly than a blink and longer than a lifetime falls outside the range of experience and emerges as a candidate for scientific inquiry. Without visual corroboration of atmospheric change, it is hard for many people to swallow results published in scientific journals. The invisibility of carbon dioxide emissions to the naked eye itself is part of the reason it has been so easy for deniers to confuse the public about dangerous man-made global warming for more than twenty years.

Keeling's invention of his own instrument for measuring the pressure of CO_2 in a gas mixture was not atypical of the some-assembly-required element of midcentury experimentation. James Lovelock, a peer who rose to prominence measuring trace amounts of atmospheric gases, has said, "In those days it was usual for scientists to make or at least design their own instruments. Most laboratories then had a workshop with metal working tools, lathes and milling machines, and scientists were expected to be able to use them . . . The greatest advantage to come from making one's own apparatus is that sometimes it is an invention whose novelty makes it years in advance of anything available in the marketplace."[15]

When Keeling reflected on his now-historic carbon dioxide measurements, he said:

> Why did I devise such an elaborate sampling strategy when my experiment didn't really require it? The reason was simply that I

was having fun. I liked designing and assembling equipment. I didn't feel under any pressure to produce a final result in a short time. It didn't occur to me that my activities and progress might soon have to be justified to the sponsoring Atomic Energy Commission. At the age of 27, the prospect of spending more time at Big Sur State Park to take suites of air and water samples instead of just a few didn't seem objectionable, even if I had to get out of a sleeping bag several times in the night. I saw myself carving out a new career in geochemistry.

I did not anticipate that the procedures established in this first experiment would be the basis for much of the research that I would pursue over the next forty-odd years.[16]

To address his actual project, which was determining the CO_2 content of streams, he needed a baseline measurement of airborne CO_2. Keeling found that the CO_2 content of air changed over the course of a day, with regular rhythm. At night the Sun's photons struck the other side of the planet. Trees shut down for the day. Carbon dioxide accumulated over vegetation at night as plants burned carbohydrates back into the air. The released gas had a pronounced higher level of carbon-12 than carbon-13 relative to neutral air samples, an indication that the plants themselves released the gas. By dawn, the air's carbon dioxide content above trees and grassy plains became inputs for the next day's photosynthetic work. By afternoon, the air mixed, showing a universal measurement of about 310 parts carbon dioxide per million molecules of air, and in regular isotopic patterns. Keeling's measurements showed the same rhythm everywhere he went, from Big Sur to the Olympic Peninsula in Washington to Arizona's altitude forests. Somehow, a conventional wisdom about atmospheric CO_2 had emerged without anyone taking precise measurements.

In 1956, Keeling sent his data on the "surprising near-constancy"[17] of CO_2 presence in air to colleagues and a U.S. Weather Bureau scientist. Soon, he found himself on a plane (first time) to Washington to argue for continuous CO_2 measurement. He hypothesized that the level was more stable than previously believed, and he wasn't sure why. He suggested detectors be placed around the world to gather samples—a call that was met by the Weather Bureau.

Keeling's results made him a respected and eventually celebrated

With university and government support, Charles David Keeling
began measuring atmospheric carbon dioxide in 1958.

leader of big science and the federal bureaucracy that supports it. Roger Revelle was director of the Scripps Institute of Oceanography. He took an interest in Keeling's data and soon offered him a position at Scripps.

Revelle's legacy looms over geochemistry. In 1957, as Keeling was scaling up his CO_2 measurements, Revelle and colleague Hans Suess calculated that oceans absorb about half the fossil fuel CO_2 emissions. Revelle wrote in conclusion,

> Human beings are now carrying out a large scale geophysical experiment of a kind that could not have happened in the past nor be reproduced in the future. Within a few centuries we are returning to the atmosphere and oceans the concentrated organic carbon stored in sedimentary rocks over hundreds of millions of years. This experiment, if adequately documented, may yield a far-reaching insight into the processes determining weather and climate. It therefore becomes of prime importance to attempt to determine the way in which carbon dioxide is partitioned between the atmosphere, the oceans, the biosphere and the lithosphere.[18]

Revelle and Keeling disagreed on a course for the first few years of CO_2 measurement. Revelle thought periodic snapshot readings of CO_2 levels at sea, in the air, and at remote locations would capture the behavior and abundance of the trace gas. Keeling pushed to continuously test the air. To do so he would acquire several $6,000 machines. In the spring of 1958 an engineer plugged in a gas analyzer at the Mauna Loa Observatory, in Hawaii. Based on previous observations on the pier at Scripps, Keeling told the Mauna Loa team to look for a carbon dioxide reading of about 313 ppm. The first reading came in within 1 ppm of that mark, a coincidence, but an auspicious start for the adventure ahead.

Today, four gas analyzers sit atop four towers at the Mauna Loa Observatory, sampling air for carbon dioxide four times an hour. The device flashes a laser of infrared light through the gas. The amount of the energy that is absorbed correlates to the amount of CO_2, with uncertainty of about 0.1 ppm.[19] Every week researchers collect gas samples from one hundred other sites around the world and determine their content of CO_2 and other gases.[20] When Keeling's first apparatus was set up at Mauna Loa, the reading was 313.4 ppm.[21] Today it has passed 383 ppm. During this indomitable rise, the Keeling curve became an icon, a simple x-y axis, charting the continuous reading of carbon dioxide levels at the Mauna Loa Observatory since March 1958. Instrumentation problems caused a short aberration in 1963 and 1964—in the midst of the first of his forty-year career of fights for funding to keep his measurements alive.[22] The Keeling curve is the oldest direct record that exists of carbon dioxide in the atmosphere. Up it goes, 2 ppm a year or so, and accelerating.

Within two years Keeling discerned seasonal variations in the CO_2 concentrations, evidence that the planet breathes to a regular rhythm. Concentration peaks in spring, after the lifeless winter. By fall, CO_2 dips after slipping into green skivvies during the growing season. The northern hemisphere's seasons dominate the cycle because most land is north of the equator, and produces more vegetation than the southern hemisphere.

As early as March 1960, Keeling could already see that fossil fuel burning might be contributing to an observed increase in CO_2. "Where data extend beyond one year, averages for the second year are higher than the first year," he wrote.[23] From year two of the experiment, through the present, the data keep moving up. By the early 1970s, Hans Suess at Scripps had detected declining relative rates of atmospheric carbon-14, the radioactive isotope used in dating materials no older than 50,000 years.[24] Fossil fuels are more than 4,000 times older than that. A smaller proportion of

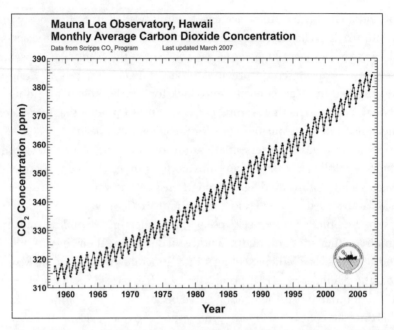

The "Keeling curve" has become an iconic image for the rise of carbon dioxide in the air and global warming.

carbon-14 would be expected if mineralized carbon were changing the composition of the atmosphere and consequently the amount of solar radiation trapped in the Earth system.[25]

The Earth's radiation budget is the flow of energy that comes in from space and departs for it. Like personal spending budgets, incoming and outgoing energy must balance for stability. Anthropogenic carbon emissions are causing the Earth to rebalance its energy budget—raising the temperature and the oceans' acidity, melting ice sheets—to accommodate the extra energy trapped in the system.

Solar energy, or radiation, approaches the Earth with an average power of 342 W (watts) per square meter—enough to light four or five old-fashioned incandescent bulbs.[26] About 30 percent of the radiation bounces off the atmosphere, clouds, ice-covered poles, snowcapped mountains, dust, aerosol particles, and lighter-colored areas of the surface. Clouds have a dual role. They trap heat between themselves and the ground. But on balance their reflective whiteness bounces light away from the surface, exerting a net cooling effect.[27] The surface absorbs the other 70 percent, maintaining the planet's 14°C (57°F) average temperature. Earth continuously sheds about 240 W per square meter. That rate should have a temperature of about −19°C (−2°F). Scientists consider this the would-be temperature, absent a greenhouse. Heat-trapping gases absorb and reradiate back to the surface enough infrared energy to raise the surface temperature it has hovered at since the end of the last ice age, 12,000 years ago.

Infrared radiation—heat—isn't powerful enough to break molecular bonds in water and carbon dioxide. Molecules absorb photons and consequently flip, toggle, stretch, and bend. The infrared light that H_2O and CO_2 block passage to is not as destructive as light of higher energy (and shorter wavelengths), such as ultraviolet and x-rays. Michael Crimmins in his ginkgolide B synthesis used a mercury light that shined a 366 nm wavelength on the intermediate molecules. That light was strong enough to break chemical bonds. Infrared light makes molecules such as water and carbon dioxide shake but not break. H_2O and CO_2 absorb energy around 15 and 12 microns, enough to jolt their electrons into higher energy levels—physics talk for absorbing heat.[28]

Two-atom molecules, such as N_2 and O_2, lack the structure to bounce, toggle, stretch, and bend the way bigger molecules do. Infrared energy moves right through them. John Tyndall knew that much 150 years ago. Water and CO_2 rotate. They also vibrate in several ways when they absorb photons. H_2O is shaped like a molecular boomerang. The oxygen atom sits at the vertex, flanked by hydrogen atoms, boomerang shaped. The molecule can stretch, a hydrogen atom drawing away from the oxygen as its twin slides toward it. The two hydrogen atoms can bend toward each other, shortening the bond angle between them. All three atoms can stretch farther away from each other than normal. Carbon dioxide molecules vibrate in several ways when they absorb heat energy. The two oxygen atoms can bend down toward each other, violating CO_2's 180-degree linearity. Carbon can toggle between the two oxygen atoms, lengthening or shortening their bond lengths. CO_2's bending vibrations absorb energy at a similar wavelength to water.[29] Sooner or later, the molecules emit their energy as heat, either back down to Earth or out into space, to begin the cycle anew.

A suite of gases warms the greenhouse. Carbon dioxide and water vapor are only the two most important. Methane, nitrous oxide, hydrofluorocarbons, perfluorocarbons, sulfur hexafluoride, and other industrial chemicals add to the problem. Many of these molecules absorb heat at 8–12 microns, just below water and CO_2.[30] Methane monitoring began in the early 1970s, after atmospheric chemists discovered a twentieth-century spike. Methane rose about 1 percent a year through the 1980s, and then stabilized in the 1990s with about 0.4 percent annual increases.[31] In a molecule-by-molecule comparison, methane is about twenty-one times more powerful a heat-trapping gas than carbon dioxide. But taking their overall atmospheric abundances into account, the warming push from methane is half as large as that from CO_2. If nations and industries don't lower CO_2, the methane level might not matter.

Other gases trap heat in the atmosphere, including nitrous oxide, the toxic product of imperfect combustion better known as laughing gas. Man-made warming gases, such as chlorofluorocarbons, are entirely new to nature. They also destroy the ozone layer, that protective layer of O_3 molecules that reduces the amount of ultraviolet light that reaches land.

Signals are pointing in the wrong direction. The Global Carbon Project is a scientific consortium founded in 2001 to fill in gaps in scientific understanding of the carbon cycle. Researchers have documented a trifecta of trends that are accelerating man-made CO_2 emissions. From 2000 to 2006 the annual growth in emissions rose from 1.3 percent to 3.3 percent. Increased coal burning has at least temporarily reversed a thirty-year trend and caused the carbon intensity of every dollar of world economic output to inch upward. (Critics of this study say three years of data does not constitute a long-term trend.[32]) That means for every unit of world output, more carbon is released into the atmosphere. In 2002 and 2003 the Mauna Loa Observatory registered the first back-to-back atmospheric CO_2 increases greater than 2 ppm.[33] Finally, carbon sinks in land and sea are filling up. Carbon has fewer places to hide.

Climates are temperamental. Even without shocks, climates can enter periods of chaos and transform the viability of ecosystems. "The paleoclimate record shouts out to us that, far from being self-stabilizing, the Earth's climate system is an ornery beast which overreacts even to small nudges," writes Wallace Broecker, a leader in modern Earth system science at Columbia's Lamont-Doherty Earth Observatory.[34] If a climate has a weak push, nonlinear change can develop seemingly out of nowhere. Anthropogenic climate change is not a weak push. Scientists call geophysical stressors on the climate forcings, or aberrations in the radiation budget.[35] Volcanoes, changes in solar energy, and variant greenhouse gases are all considered forcings. The IPCC concludes in its fourth assessment report that anthropogenic radiative forcings are adding another 1.6 W per square meter to the Earth's energy budget. Another 0.6°C (1°F) of warming is already built into the system, even if our cars and power plants stopped burning carbon tomorrow.

When Svante Arrhenius conjectured what a doubling of atmospheric carbon might do, he set a precedent that eventually turned into widespread scientific convention. Climate sensitivity is the common yardstick for predicting the effect of increasing atmospheric carbon. Scientists define it as the global average temperature rise from a doubling of preindustrial CO_2 levels. CO_2 doubling will occur when the trace gas reaches 560 ppm, twice the 280 ppm before industrialization. We're currently in the mid-380s. At the current rate of fossil fuel consumption, that Rubicon might be crossed by 2050. Another informative measurement is called the carbon dioxide equivalent (CO_2e). This figure includes the warming effects of the other greenhouse gases. The atmosphere contains about 430 ppm CO_2e at this writing. Current modeling shows that the probable climate sensitivity to a doubling of CO_2 falls between 2°C and 4.5°C. The average predicted increase is 3°C, according to the fourth assessment report of the Intergovernmental Panel on Climate Change.[36] If successful, the Kyoto Protocol would have stabilized CO_2 emissions at 1990 levels. Emissions have risen 35 percent since then, a bad sign.

From a human perspective, many things are predicted to go wrong with the Earth this century and next. Some scientists already recognize humanity's impact on biodiversity as the sixth major extinction since the Cambrian period. The Southern Ocean (Antarctic Ocean) has already shown a decreased appetite for dissolving carbon. The land ecosystems that absorb atmospheric carbon will probably peak before 2050 and possibly reverse, becoming part of the problem rather than part of the solution. A third of plant and animal species may face extinction if temperature rises past 1.5°C–2.5°C. Up to 250 million Africans may face climate-change-related water shortages by 2020, with consequent implications for growing food. Drought will visit Australia, southern Europe, and the southwestern United States. Heat deaths, species migrations, and disease spread as the twenty-first century progresses. "It is clear that the future impacts of climate change are dependent not only on the rate of climate change, but also on the future social, economic and technological state of the world," the IPCC authors write.[37]

James Hansen, director of NASA's Goddard Institute for Space Studies, says that an Earth 2°C warmer than preindustrialization would send the climate past the point of no return: oceans (at least) several meters higher, extinction of a third to half of animals and plants, the decimation or collapse of ecosystems. "These consequences are no longer speculative climate model results," he has written. "Our best estimates for expected climate impacts are based on evidence from prior climate changes in the Earth's history and on recent observed climate trends."[38]

The changes wrought by global warming threaten the extent and potential of modernity. Only nuclear war or a drug-resistant pathogen can compete with global warming as civilization's greatest threat. The simplest and most scientifically sound description of anthropogenic climate change can be summarized in two sentences:

1. Temperature and atmospheric carbon are coupled on every geologic timescale, usually with temperature leading—but not since industrialization.

2. Humans are burning carbon minerals into atmospheric gas at least one hundred times faster than the Earth's usual rate, heating and transforming the planet.

Together these two statements challenge the current meaning of the phrase "geologic timescale." Human speed has crunched the geologic timescale into half a century. Events that typically unfold over many thousands or millions of years have begun to occur within a human life span. Anthropogenic global warming erases the line between the biological and the geological timescale as vividly as anything else humans do on the planet.

Robert Berner of Yale emphasizes the difference between the long- and short-term carbon cycles. The short-term carbon cycle lasts fractions of a second to dozens or hundreds of years. The long-term carbon cycle covers tens of millions of years, and includes the lithosphere, the shell of rock that separates loose soil and ocean sediment from the mantle. The lithosphere is the domain of plate tectonics, the centimeters per year slide of the twelve continental plates into, beneath, atop, or along each other. It's the domain of weathering and sediment, ossified by heat and pressure into rock. Rain and rivers bearing carbonic acid draw minerals from rock and wash them to sea. Rock is subducted into the mantle. Silicate rock weathering is the major contributor to carbon cycling over geological time. But rocks rich in organic carbon weather, too—about one hundred times more slowly than they have since industrial fires ignited.

Life has always driven and been driven by geology. The flow of carbon through living things entwines evolution with the inanimate forces of nature. But there is no evidence before now to suggest that biology has ever accelerated the long-term carbon cycle onto a short-term path. Nothing other than meteorite impacts have changed geology as quickly as humanity. Industry is a powerful new path of interaction between life and geology. "We *are* plate tectonics!" said Scott Wing, director of the Smithsonian Institution's paleobotany division, reaching for a way to characterize the power of human-induced change.[39]

The rate of burning is all the more astounding given how long it took to put those fuels in place. Oil comes from decaying biomaterials, the organic rain of coccoliths, diatoms, copepod fecal pellets, foraminifera, pollen, spores, and other once-living things. Heat, pressure, and time degrade them into straight-chain hydrocarbons, aromatic rings, carbonaceous chicken wire, and every variety of other black gunk.[40] Humans in a century and a half have burned as much as half the recoverable oil that took tens of millions of years to bleed out of their carbon-rich source rocks.

The late German writer W. G. Sebald hides a perspicacious summary of humanity's effect on the Earth in his 1995 novel, *The Rings of Saturn*. The book's peripatetic narrator strolls the shores of England, ruminating on seascapes and history. The ravaged forests of Dunwich Heath in southeast England remind him of the centrality of burning in the rise of nations:

> It's not for nothing that Brazil owes its name to the French word for charcoal. Our spread over the earth was fuelled by reducing the higher species of vegetation to charcoal, by incessantly burning whatever would burn. From the first smouldering taper to the elegant lanterns whose light reverberated around eighteenth-century courtyards and from the mild radiance of these lanterns

to the unearthly glow of the sodium lamps that line the Belgian motorways, it has all been combustion. Combustion is the hidden principle behind every artefact we create. The making of a fish-hook, manufacture of a china cup, or production of a television programme, all depend on the same process of combustion. Like our bodies and like our desires, the machines we have devised are possessed of a heart which is slowly reduced to embers. From the earliest times, human civilization has been no more than a strange luminescence growing more intense by the hour, of which no one can say when it will begin to wane and when it will fade away. For the time being, our cities still shine through the night, and the fires still spread.[41]

By letting the fires spread—yielding to an age of ever-larger candles stoked by the invisible hand of economic competition—we as individuals and as a society, as nations and as a species are deciding that our lifestyle is more important than its continuity and a vast percentage of species on Earth. Humans did not develop in a vacuum, but our actions suggest we believe we exist in one.

But it's possible to slow down our impact on the carbon cycle without sacrificing our industrial fire, if we move fast. Technological investment in new energy and materials industries could remake the way we make things. To do so will require a more open discussion about fundamental moral choices we make as a society, particularly ones obscured in the professional language of science and economics. To the lay observer, there's much obscurity in the climate crisis, whether it's the jargon of scientists or the jargon of economists. Through this obscurity, and through the cracks between professional communities who don't always communicate as well as one might hope falls a sobering fact. What scientists describe as well-beyond their danger zone, economists and politicians treat as the bottom of the potentially achievable.

To gauge the distance between scientists and economists, look at how they each treat two numbers: 560 ppm and 550 ppm. The former number represents a doubling of preindustrial carbon dioxide levels, and all of its potentially disastrous consequences. It's likely that temperature is well past the thumbnail 2°C tipping point, beyond which the climate and the feedbacks that accelerate change will respond in ways that are even more difficult to predict. Yet for the economic mainstream, the latter number is a medium-to-low-end stabilization target. That's a tragic paradox. The upper limit of scientific fear is a CO_2 level unnervingly below economic and political interest.

The Intergovernmental Panel on Climate Change is divided into three working groups. Working group I reports on the physical science of global warming. Working group II studies impacts, adaptation and vulnerability. And working group III models the potential costs of mitigation. In some scenarios, the IPCC scientists model impacts at 700 or 800 ppm, and economists entertain scenarios to value them. The problem is that allowing ourselves to model carbon emissions, temperature change, impacts, impacts caused by the impacts, and finally potential economic injury of mitigation at levels reaching 700 or 800 ppm may be like talking about the earning potential of a patient diagnosed late with a terminal illness. Chronic illness is to personal income what global warming is to world economic output. The atmosphere may not have contained 700 ppm in millions or tens of millions of years. Economics was hard when climate conditions were stable; unstable, unpredictable climate conditions make economic estimates one step removed from the unknown. What we do know is frightening enough: CO_2 emissions may be advancing along a pathway that is faster than the scenario IPCC scientists considered too pessimistic to include in the fourth assessment report.

When it comes to predicting the future, economists have drawn the short straw. That's because they must model human mass behavior on top of the scientists' models of the Earth system. The scientists' job is difficult but straightforward. A common research question for them is, how much is the temperature likely to rise if the atmosphere's CO_2 content doubles from its 280 ppm level, to 560 ppm? They home in on a probable range, given physical and proxy evidence, and their understanding of Earth physics and chemistry. The IPCC's likely range for the climate's sensitivity to a doubling of CO_2 is between 2°C and 4.5°C, and most likely about 3°C. From these studies, scientists then predict the potential magnitude of the impacts that may result. Projecting impacts is a monstrous task, because the number of things that can go wrong—and the rates and magnitudes of their doing so—are so varied.

The economists then model how reducing carbon emissions might affect future world economic output. The trouble is, all these things become increasingly difficult, each prediction building atop the uncertainties in the scientific predictions. We know a lot about the Earth system, how carbon flows through it and warms it. It seems like modeling human behavior should be easier than modeling the Earth's global carbon cycle. It's not. Energy and material flow through the Earth system with more regularity than people drive an economy. In many ways, human behavior is more complicated than the Earth's. The planet has no will.

Predicting possible futures depends on both climate and economic models, and their interaction. One tool for thinking about economic impacts is through what economists call the social cost of carbon, an estimate of the impact of one ton of carbon emissions—or a ton of averted emissions—at some point in the future. Modelers calculate the social cost of carbon by looking at how long those tons hang in the atmosphere, the warming they impose, and the potential feedbacks they trigger for even more warming and how to value it in the present. Carbon's residence time is a moving target. The more carbon in the atmosphere the longer it stays. This means that economists need to know how long carbon resides in the atmosphere, the impacts, the value of those impacts, and also how much they might be valued at the time of the carbon emission. Atmospheric chemists have a hard enough time, let alone the economists who must run models on top of models. And as it is, Richard Tol of Hamburg University has written that projecting the economic impacts of climate change is still immature.[42]

Evaluating the potential impacts of global warming is a monumental task. Scientists must estimate everything that might go wrong, the feedback loops within everything that might go wrong, and at what rate and magnitude everything potentially goes wrong together. Rising temperatures evaporate more water, which raises the temperature further; ice sheets melt and reduce the Earth's reflective surface area, inviting in more heat; soils release stored carbon; oceans absorb as much carbon as they can, leaving more to accumulate in the atmosphere. Jim Hansen's statement in the epigraph to this chapter is not forward looking. He says that the Holocene is about to end. No one can say with certainty how quickly the Earth will change from here. Informally, this next phase of Earth history has already taken on a new name, the Anthropocene, or human epoch.

So, here are two tasks handed the economists. First, how do you value future events when you don't know what is going to happen? "Every cost-benefit analysis is an exercise in uncertainty," Martin Weitzman of Harvard has written.[43] If global warming is the biggest externalized cost the world has ever seen, he wrote, estimates of its impact may be the biggest exercise in subjectivity economists have ever undertaken. Second, how do you value events even when you know they are likely to happen? Ruth Greenspan Bell, a consultant on climate issues in Washington, D.C., asks economists the logic of making sure our progeny have a healthy fiscal picture in the face of Earth's changing or declining life-support systems: "How will richer future generations replace the Greenland ice sheet?"[44]

Climate and economic modeling starts to take on a chicken-and-egg feel after a while. Projected carbon dioxide levels depend on economic models of future growth in world output and associated carbon-base energy output. Economists have to measure the costs of preventing or recovering from damages that no one can say with specificity will occur.

The climate conversation turned a corner in 2006, partly because these considerations became the subject of more public debate. The scientific debate about whether man-made emissions were changing the climate ended years before, though most politicians, industrial leaders, and media failed to reflect that. The persistent work of the Intergovernmental Panel on Climate Change, scientists the world over, and communicators—most effectively Al Gore—made the most potent frontal assault on the most potent disinformation campaigns.

In 2006, the conversation shifted from science to economics, specifically the influence that some economists have had in climate debates to this point, and importantly, the assumptions they make in charting the future. A British government study led by former World Bank chief economist Nicholas Stern broke with mainstream economic thought and forecast monumental economic hardship the more the climate changes. The report recommends immediate and overwhelming efforts to reduce carbon emissions. It caused controversy in part for its truly frightening conclusions. The authors wrote: "The Review estimates that if we don't act, the overall costs and risks of climate change will be equivalent to losing at least 5% of global GDP each year, now and forever. If a wider range of risks and impacts is taken into account, the estimates of damage could rise to 20% of GDP or more."[45]

In reaching these conclusions, the Stern Review laid bare the device by which economists evaluate the costs and benefits of investment now versus the future. It's more art than science, but art with a lot of calculus. Climate change polemics in economics now fly over who will or should or can bear the greater burden, ourselves or our posterity. Much debate centers on the so-called discount rate, an expression of how much less a dollar is worth in the future than today. The variables chosen by an economist are plugged into a standard equation, and out pops an analysis of how the value of spending changes over a given time period, and whether the future social benefit of spending is worth the current expense.

Generally, people would rather have a dollar now instead of a dollar in a hundred years. In other words, a dollar a hundred years from now is worth less than a dollar today, perhaps due to the fact that we see ourselves getting richer, or perhaps we just want the dollar now—since we personally will not be here in one hundred years. The Stern Review uses a very low discount rate, which means that the dollar today and the dollar in 2100 are virtually coupled. Its authors view future generations as having virtually equivalent importance as our own. A higher discount rate gives less weight to the future, assuming that a dollar tomorrow is worth much less than one today. In thinking about the unforeseeable future—a century ahead—changes in the discount rate have enormous influence on the projected value of the dollar and therefore on the question of whether we should invest now to prevent or slow catastrophic change or save dollars, so that future generations will have enough money to recover from it.

An economist's choice for inputs into the so-called Ramsey equation has enormous repercussions for the answers they get and therefore the climate-economic policies they advise. In one IPCC scenario, if the discount rate is 3 percent, the social cost of carbon comes out to $62 per ton of carbon emissions. That's low enough that a company might be willing to buy credits rather than investing in low-carbon technology. Discounting at 1 percent yields $165 per ton. A 0 percent rate bumps the social cost of carbon to $1,610 per ton—high enough to ensure that deploying virtually any new technology or cutting emissions in any way will be cheaper than buying pollution credits in a carbon market.[46] This may seem obscure, but policies that will change the world, and how we and our children live, are influenced in no small part from economists' discount-rate assumptions. Lawmakers have to consider these assumptions, and their complex moral implications, in deciding how to respond to global warming. Richard Tol summarized the matter succinctly, in a paper about the social cost of carbon. Stern Review–like analyses "may be morally preferable, but are clearly out of line with common practice."[47] If common practice is not morally preferable, then how is it morally defensible?

As long as we are pegged to an economic orthodoxy that equates well-being with per capita income—and we have no competitive alternative—we are unlikely to address the fundamental drivers of climate change: materialism, crass commercialism, and waste made easy by cheap, plentiful fossil fuels. Fire bound together human communities in the many thousand years before our physiology even reached its current form. The fire has gotten out of hand since then, requiring community cooperation on a global scale, cooperation that surmounts rending issues of equity and justice, argues John Gowdy of Rensselaer Polytechnic Institute.[48] That's a powerful goal and a hard sell.

Weaning civilization from the fuels that enable it, without disrupting civilization, is the most difficult civic works project ever undertaken—much harder than growing civilization in the first place. Hans Bethe said about his 1939 findings about how a star shines that it "eats its carbon and has it, too." In a radically, different context, that is what we are trying to do with our energy system. And we're not doing a very good job. Robert Socolow of Princeton and colleagues developed a plausible way to change slowly—if we start immediately. It very quickly became a widely accepted framework for thinking about how energy transformation should happen. Socolow and Princeton colleague Stephen Pacala proposed the paradigm of "stabilization wedges" as a way to conceptualize change—new technologies, greater efficiency, and other tactics. To stabilize emissions at 500 ppm, they observe that carbon emissions would have to be held to 7 billion tons a year between 2004 and 2054. Unfortunately, at this writing, emissions should pass 10 billion tons in 2007, and the rate of acceleration is accelerating, mostly because of China's ascension in the world economy.[49]

For a long time, most signals have suggested modern economies need to get out of the CO_2-making businesses. People have manufactured CO_2 industrially longer than we have known what it was. Herbert Hoover knew in 1921, when as Warren Harding's commerce secretary he told the Synthetic Organic Chemical Manufacturers Association that the fires of industry wasted the carbon blown out smokestacks. "The very coke oven today that is not recovering its by-products, turning its by-products into the air, is turning a loss that can never be recovered. Your industries are the industries that take these derivatives and turn them to account . . . If we are going to maintain our own world, we must turn all these waste factors into something productive, and an industry that is almost wholly founded on the recovery of those wastes naturally is worth cultivation and encouragement, not only by the country but by the government it-self."[50]

Hoover's is still a far-off dream. No deployable synthetic technology exists, or appears on the horizon, that can scrub CO_2 from the air and turn it into something useful, like plastic or broccoli. And companies have been resistant to paying for the airborne equivalent of leaving ashes in a hearth. They're as used to it being free as breathing. The notion of paying to blow carbon out smokestacks long caused reactions as if the government charged a nickel for every exhalation. That's charging.

With public polemics doubting the main facts of global warming across the Rubicon, the new battleground is over money and how to build the extraordinarily complicated and unified global system that must be put in place to monetize and monitor national and industrial greenhouse gas emissions and ratchet them down. The winds of warming reached Washington, D.C., only recently, but they blew strongly off the Potomac, into the weight of Earth's history—up the Precambrian marble of the Capitol's eastern facade and the Ordovician period granite steps at its western face, the Cretaceous sandstone of its Rotunda. At this writing, the main achievement of the White House on climate is to convene in a build-ing whose whiteness reflects 240 W per square meter from the Earth's surface back into the sky.

In a short period of time, humanity has gone from an influential species, to the most powerful driver of evolutionary and geological change on the planet—more powerful than plate tectonics, silicate rock weathering, solar hiccups, or orbital perturbations. Some scientists, ama-teur astronomers, and Hollywood filmmakers look fearfully to the skies for civilization-ending bolides. They should look inward. We are the meteor.

Industrial energy policy is a biogeochemical force and should be thought of as a cousin of earthquakes, volcanoes, pandemic disease, erosion, and other phenomena that shape the face of the Earth. In the 1920s, Russian geologist Vladimir Vernadsky saw the wreckage and carnage of World War I and concluded that history and geology are synonymous. "From the naturalist's point of view (and I think, from the historian's) it's possible and necessary to look at historic events of such power as a single, significant geological process, not just a historical one."[51] Vernadsky's comments appear in an essay about the Noösphere, from the Greek for memory, his name for the human epoch. "In it, for the first time, humanity becomes a powerful geological force."

Agriculture is responsible for about a quarter of the carbon that terrestrial plants patch into their tissue every year, an unprecedented impact by a single species, particularly given the geological instant we've taken to do it.[52] Volcanoes and weathering typically reshape the Earth's face. Today, industry may sculpt the Earth every year more than they do.[53] Human activity has transformed virtually every ecosystem on the planet. Human technology accelerates evolution by altering the ecosystems of bacteria, crop-loving insects, influenza viruses, fisheries, and really just about everything else on the planet's surface and much beneath it. Bacteria quickly evolve immunity to antibiotics, swapping genetic material through horizontal gene transfer. Thin, long fish escape trawlers' nets and live to populate the ocean with offspring slinkier than captured fish. Farmers buy more and more pesticide every year, as bugs evolve their way around them. Technology imposes unnatural selection like nothing else.[54]

Perhaps a generation or two from now, leaders will look back and wonder how American universities graduated class after class into an economy run on carbon-mineral fuels, without requiring any of them to take introduction to geochemistry courses, or why an environmental "movement" sprung from toxic chemical pollution produced so many lawyers and so few chemical engineers.[55] The way things are going, students are showing more interest in the Earth system and the chemicals we invent for it, but perhaps not as much as the gee-whiz revolution in biology that could provide some important tools to address the causes of global warming. Science helped create and diagnose the planet's fever, and it is our only hope to slow its acceleration.

David Rind of NASA's Goddard Institute for Space Studies is less than sanguine about modernity's likelihood to change. "Nobody's really going to sacrifice the present for the sake of the future, no matter how much they say it," Rind says. "People who are born in 2050 will have a lower quality of life, but they won't know any better. The sun will still shine and people will live their lives, but it will be a different world and nothing will change that."[56] The real goal should be to prove Rind wrong. If he's right—which so far he is—forty years from now some of those Americans living a lower quality of life may wonder about the generation that knew they had it all and set decline in motion, unfazed. Ours will be a generation of narcissism. Each one of us is competitive with Nero, who fiddled while Rome burned, frittering away time on bread and circuses. Each of us is Tsar Nicholas II, whose attention to traumatic personal circumstances—an ill son—distracted him from his empire, crumbling toward what would become the Soviet dictatorship. The climate debate stands at an analogous point. There is still some hope that industrialized nations can transfer civilization on to an energy system that will not scorch the Earth. Hope springs eternal. Opportunities pass.

11 | Instructions Not Included
The Potential of Biological Fuels

In a sense, the fossil fuels are a onetime gift that lifted us up from subsistence agriculture and eventually should lead us to a future based on renewable resources.
— Kenneth Deffeyes

The solution to climate change is to stop burning carbon minerals into atmospheric gas and removing forests. This is much easier said than done, as evidenced by our two-decade-long, still-sputtering start.

To set this goal in broader terms, industry needs to find a way to live inside the biosphere, inside the short-term carbon cycle, where energy resources are renewable, where energy is stored in and moves among living things—not dead ones who survive only in hydrocarbon fossil. Until now, this has been an impossible task for three reasons: First, elected officials have not shown leadership and spent enough money on research and deployment of carbon-free technologies, both those under development and those ready for widespread use. Second, these technologies have accordingly been unable to reach scale. And third, scientists are only beginning to understand biochemistry enough to find possible answers to our energy and climate crises within it.

Biotechnology research is pointing in a direction that may lead to the gradual harnessing of life-processed carbon for industrial energy. These are dreamy, far-off answers, but the most logical ones to sustain everything we've built—and we're inching toward them. We'll burn coal and oil for the foreseeable future, probably the unforeseeable future, but potentially mitigating their hazardous CO_2 waste. Already, several infant companies are attempting to scale up production of microbes genetically engineered to augment fermentation processes that lead to transportation fuels either identical to those derived from petroleum or more powerful than conventional biofuels. These fuels can supply existing motor vehicles, without burning "long-term" carbon into the "short-term" cycle and without requiring farmers to grow crops for fuel instead of food. But they can't do it yet. The U.S. government funds some research into the genetic engineering of organisms to produce industrial fuels. Some of the efforts target more efficient ways of making ethanol from cellulose, or making it from lignin—the tough material of plants and trees—instead of flimsier cellulose. Bacterial fuel-cell research might lead to microbes that generate electricity from many sources of carbon, including sewage.[1]

We can't stop burning carbon, but we can limit it and strive to change where it comes from. Useful applications may come from a nascent field called synthetic biology, the invention of biological machines and processes by altering natural genomes and metabolisms—and by creating new ones. Synthetic biology is a rising union of research and researchers—potentially a second origin of life, with all its promise and hazards. It's a young field, but one that sets its sights on how to grow, not mine, industrial carbon fuels. This chapter chronicles the goals and launch phase of this intriguing union of the natural and the unnatural.

The history of *Homo sapiens* is written in combustion, in biochemical and industrial pathways that recover energy from carbon bonds. The basic principle is the same for each—breaking bonds, harnessing the energy, and remaking lower-energy ones. Before fossil fuels came along, the power plants of every animal cell—mitochondria—drove muscle and therefore civilization by burning glucose derivatives and transferring their energy into biology's universal fuel, ATP. People, oxen, draft horses, and sled dogs continuously use and replenish it in the course of tilling fields, drawing carriages, or mushing. Mitochondria have their own history, their own identity, entwined with their host cells but not coincident: they have their own DNA. Mitochondrial DNA (mtDNA) has a feature that distinguishes it from nuclear DNA, the material wound inside our every cellular nucleus. Nuclear DNA in humans winds into twenty-three pairs of chromosomes, one complete set from each of our parents. The genome is a team effort, with both parents contributing. The same can't be said of mitochondria. Their DNA descends directly from the mother, which she took from her mother, and so on.

In the 1980s, scientists at the University of California–Berkeley studied mtDNA in women from across the globe. The reading of DNA base pairs is called sequencing. Mutations in mtDNA occur at a regular rate. Counting backward, Allan Wilson and his team concluded that every human being on the planet descended from an African woman who lived more than 150,000 years ago, our common Earth mother, whom they dubbed Mitochondrial Eve.

Eve gained a husband of sorts several years later. Sex is determined at fertilization, when a sperm (from a human, *Ginkgo*, what have you) carrying either an X or Y chromosome penetrates the egg. An X joins its sister in the egg, and produces an XX combination—a girl. A Y joins the mother's X chromosome to produce an XY combination—a boy. Just as mtDNA passes through the generations along a matrilinear path, the Y chromosome transfers from father to son.[2] Y-Adam, the common ancestor of human males, similarly comes from a band of African hunter-gatherers.[3] Neither mtDNA nor the Y chromosome undergoes change that genetic recombination brings to nuclear DNA.

Scientists understand the dissemination of a species through the ages based on comparisons of specific regions of nuclear DNA coding and mitochondrial DNA. These studies belong to the field of phylogeography. For example, brown bears colonized Europe from the Iberian Peninsula, in the southeast, and from the Caucasus south of Russia. The two lineages made their way north, eventually meeting up in Sweden.[4]

Human genetic material, nuclear or mitochondrial DNA, can be sequenced and backtracked in the same way, as have bears, hedgehogs, grasshoppers, sparrows, and others. The Genographic Project, launched in 2005, is an effort by the National Geographic Society, IBM, FamilyTree USA, and Arizona Research Labs to create a public research database of sampled mtDNA, a "real-time anthropological genetics study."[5] The team published its first peer-reviewed paper in June 2007, by which time more than 188,000 people had signed on to the project.[6]

By calculating the genetic distance among subjects, the scientists can estimate the waves of exodus from Africa, which began in earnest about

50,000 years ago.[7] The Middle East colonized 45,000 years ago. Europe fell 35,000 years ago. An ice bridge across the northern Pacific late in the last ice age dropped *Homo sapiens* in the Americas. In 50,000 years the human population spread from its African home to today, when humans live on land everywhere from North Pole to South. All along this journey, people reproduced and genes recombined, amplifying what few differences there are among us within the genetic code (about one from every thousand base pairs). That's a remarkable fact. In 50,000 years, humans went from a band of two thousand fire-wielding bipeds to the most influential force for evolutionary and geological change on the planet.

The Genographic Project and other phylogeography studies take advantage of the steep advance of genetic technology and precipitous decline in the price of sequencing DNA. Its name hearkens to a previous, more famous enterprise, the supposed alpha and omega of biology: the Human Genome Project, an initiative started in 1990 by the U.S. Department of Energy and the National Institutes of Health, and the British government. Their goal was to map the human genome within fifteen years. They did it in thirteen, eventually under competitive pressure from the iconoclastic and entrepreneurial Craig Venter. The company he founded as a competitor, Celera Genomics, deployed faster sequencing technologies in the mid-1990s and caught up with and in some ways surpassed the interagency, international behemoth. The race to transcribe the genome from a chemical script into a chemistrese-derived alphabet (A, T, G, C) became an arms race between Venter's upstart company and the U.S. initiative.

Improvements in the speed of gene reading are reminiscent of the rapid growth of the number of transistors used in computing. Moore's law describes semiannual to biannual doubling of the amount of transistors that fit on an integrated circuit.[8] Hundreds of millions of transistors fit on modern chips—compared with fifty when Intel's Gordon Moore made the prediction in 1965. Even when silicon technologies begin to reach their physical limits, carbon-based successors wait in the wings. Carbon nanotubes are cylindrical graphene sheets—rolled up molecular chicken wire—that scientists have built into experimental transistors.

There are many parallels between computing and biology. Cyberpunk and science fiction authors have long conceived futures in which the "meatworld" of our bodies fades into a digital oblivion. Code is code, regardless of medium, natural or unnatural. Pure information moves through species or computers by analogous methods of replication and selection, whether you call them "memes" or "useful knowledge" or "snippets." Computers churn a binary code of zeros and ones. A four-letter chemical alphabet— A and T, C, and G—stores biological information. The English language is a twenty-six-character code (not including parentheses, spaces, commas, periods, and the like). The human genome is about three gigabytes long, the equivalent of about ten copies of Leo Tolstoy's *War and Peace* (without spaces, and in Russian).

The funny thing about information is that it seems to behave similarly, regardless of the code it's inscribed in or the medium it's expressed in. That's what allows us to talk about DNA as a replicator and to scan the human experience to find analogous mechanisms by which information emerges, mutates, survives, or dies. The notion that code is code, information is information, underwrites the ability of computers to mimic evolutionary behavior. For decades software has grown more sophisticated at introducing mutations into code and selecting the best optimized "offspring." Genetic algorithms are programs that set up rival solutions to a problem—grocery store layouts, battleship weaknesses, aircraft-engine designs. The most desirable parts of each solution survive into the next generation of solutions, just as genes do in biology. Solutions "mate" with each other until designers feel the computer has generated an optimal scenario. Similarly, computational analogies pervade molecular biology. If evolution is reduced to a description of how structured information behaves in dynamic systems, one could say organisms "run" DNA "software" within the "hardware" of cells' membranes, cytoplasm, and proteins.

Today, lab machines perform most of the sequencing work. They cull desired DNA from cells. Bacterial colonies "amplify" the samples, which means they grow the desired snippets into quantities sufficient to machine read; or automatic thermocycling PCR (polymer-chain reaction) machines can amplify desired sequences billions of times, in an afternoon. The DNA then must be removed from the growth colonies and fed into actual sequencers. Speed-reading DNA has accelerated faster than Moore's law since the mid-1990s. The number of base pairs one person can sequence per day, using several machines at once, has leaped more than five hundred times, and is doubling every two years. The cost of sequencing a base pair was $10 in 1990—and $0.10 in 2006.[9] As a result, not a week goes by without another genome falling, its code published on the Internet for all to see: sea urchins, mustard plants, mosquitoes, honeybees, and never-ending representatives of the microbial world.

The time needed to solve a protein's structure is declining, too, despite the complexity of these macromolecules. Scientists affiliated with the Howard Hughes Medical Institute used the power of the Internet and distributed computing to correctly predict how a specific chain of 112 amino acids folds into a protein. More than 150,000 computer users around the world volunteered computing power, collectively searching the likeliest structure the chain should snap into.[10]

By 2020, sequencing technology could make the thousand-dollar human genome a reality. That's the soft target for when technological advances might bleed over into personalized genomic medicine. Even if personalized sequencing were practical and affordable today, biology and medicine have not sufficiently advanced to take advantage of the information, the true sign of genetic medicine's infancy. The personal genome today is an occasional novelty where it isn't an object of study. In the future, patients with access to genomic medicine will have to decide how much they really want to know about themselves. When James Watson, co-solver of DNA's double helix, was presented with a readout of his own genome in 2007, he requested not to be told if he carries a version of the *APOE4* gene, which is linked to susceptibility to Alzheimer's disease. A Branford, Connecticut, company called 454 and Baylor University completed the project in two months and for less than $1 million. That's a significant change from the Human Genome Project, which cost about $300 million.[11]

The Human Genome Project was a monumental achievement of science, and has been chronicled as such, but it's only the very beginning. James Collins of Boston University describes the human genome readout as a parts list lacking an instruction manual.[12] Imagine trying to make an English-language movie out of *War and Peace* given only the Russian letters the book is written in, and assuming you don't speak Russian.

Sequencing technology will continue to improve but is really only a first step. Its price decline is not an apples-to-apples comparison to Moore's law. Shrinking transistors to fit more on a chip creates a product that manufacturers then assemble into a computer. It boosts power and memory. DNA itself only provides memory. A sequenced genome is a parts list for machines that people don't know how to assemble from scratch. There's a big difference between having a DVD player that is read-only and one that is writable. A similar distinction applies here. Sequencing is read-only memory. Scientists cannot create their own genetic DVD player—living cells to play them. DNA *synthesis* allows scientists to re-create genetic snippets—or create new ones. This technology allows labs to write in DNA, letter for letter. Synthesis technology has grown at least as rapidly as gene reading.

Firms engaged in DNA synthesis generally fall into two groups, those that make DNA snippets, usually less than 200 bp in length, and those that produce genes larger than 200 bp that scientists can splice into a genome. The average price per base pair has fallen from about $30 to about $0.20 in the last decade or so. Synthesizing gene-length DNA molecules currently costs about five times more per base pair.[13] Advances in error correction have improved the reliability of synthesis. New technologies let slip just one error in ten thousand. Of course, DNA in living systems make just one in a billion errors, but give engineers a break. They have been scaling up the process for ten years. Life had a 4-billion-year head start. The laboratory process is much slower, too, adding one base pair at the rate of one every five minutes. Some bacteria can synthesize their own DNA at the rate of about 500 bp per second.[14]

The complexity of biological systems, even the simplest cell, is a machine that eludes complete scientific understanding, let alone reproduction. But scientists know enough to create simple devices, demonstration projects on the road to fuels and medicines made by synthetic living things. An early example shows just how simple. Tim Gardner, then a graduate student at Boston University, James Collins, and Boston University biologist Charles Cantor built a genetic switch composed of two genes, only one of which can be expressed at time. Some outside influence, either temperature or chemical, changes conditions, turning off one of the genes and turning on the other. The simple one-bit computer is a far cry from silicon transistors, but the bacterial toggle switch was a conceptual and experimental breakthrough.[15]

If a songwriter needed to build an MP3 player every time he wrote a song, he wouldn't spend a lot of time at the piano. Yet that's basically what biologists are up against. Synthetic biology is about where car making was around the turn of the twentieth century. Individual shops had to make their own parts (DNA sequences) from scratch. They have learned that DNA synthesis is a time-consuming process better farmed out to manufacturers of standard parts. For engineered biological systems to flourish, scientists will need to outsource DNA synthesis to commodity service labs. As DNA synthesis technology becomes cheaper and ubiquitous, synthetic biologists will be able to rely on companies to read and write DNA, so they can focus time and money on what they really want to do, design novel or improved biological systems. Designing and synthesizing DNA parts is the first step in building new biological devices.

Genetic engineering emerged in the early 1970s when recombinant DNA—the splicing of genes into bacterial genomes to grow proteins such as human growth hormone or insulin. By the time the Human Genome Project was completed, actual engineers noticed that biological engineering and genetic engineering really aren't engineering at all. Engineers figure out how to design and fix distinct parts together into systems that perform some desirable function. Applying this paradigm to cells, "engineers could make 'wet' biological versions of electric circuits and rewire existing organisms just as they might tinker with a radio," according to Collins.

To that end the BioBricks Foundation, a consortium of MIT, Harvard, and University of California–San Francisco professors, set up a public repository for "standard biological parts," free sequence data that researchers can search through and order, as if from a catalog. Labs can order standard parts and pay only for their synthesis. A public domain catalog circumvents the need for licensing fees or litigation costs over the right to use a patented sequence. BioBricks are an homage to Lego building blocks, a serious plaything of MIT's Tom Knight, a founder of the project.

The BioBricks Foundation's earliest success was in engaging young scientists in the field's development. The first International Genetically Engineered Machine competition (iGem) was held in 2003. The next year, a team from the University of Texas–Austin gave synthetic biology its first gee-whiz creative kick. They spliced BioBricks together into a kind of living photographic paper—E. coli bacteria reprogrammed with cyanobacteria genes to react to light.

Safe strains of E. coli, long the cellular biologist's guinea pig, are not light sensitive. E. coli strains delivered through infected meats sit in the dark of intestines and wreak havoc where the Sun doesn't shine. To make E. coli see light, the Austin team spliced into an E. coli chassis two cyanobacteria genes that order the production of light receptors, called phycocyanobilins. They were also wired to emit a dark pigment in the absence of light. So, under darkness, the bacterial lawn emits the dark pigment. When light shines, phycocyanobilin antennae rattle, and tell a DNA activator to turn off the pigment-making gene. Whatever image is projected onto the lawn remains there. The dark parts of the image do not shut off dark-pigment production, so those parts of the bacterial lawn remain dark. Light turns off the pigment maker. Light areas remain light. Dark areas remain dark. The bacteria are sensitive enough that in-between areas are printed in the E. coli equivalent of a gray scale.[16] The result is a living picture. The UT Austin team shined the phrase, "Hello, World!" onto their bacterial lawn, a traditional phrase in the debut of software or computers, dating to the 1970s.[17]

To create their photogenic E. coli, the students did not culture cyanobacteria, sequence its genome, isolate the genes involved in phycocyanobilin production, and make them in a DNA synthesizer. They identified them on the Internet at http://www.biobricks.org/, and a UCSF laboratory offered copies of the parts they needed. BBa_I15010 seems like a dreadful part number to have to remember, but it beats having to determine and painstakingly reconstruct over several months the DNA makeup of the gene: atggccaccaccgtacaactcagcga . . . [2,200 bases omitted] . . . gaagggtaataa.[18]

Bonding DNA into biological parts is the first of two rungs of the conceptual ladder, which is borrowed from electrical engineering. Scientists assemble parts into devices, which perform some kind of specific function. Devices can be spliced together into novel systems that can achieve tasks useful to their human inventors—such as bacteria that glow in the dark when a deadly toxin is present, store memory, or grow antimalaria drugs.

The second rung is standardizing parts and putting them together, especially demanding tasks on a scale of 2 nm, the width of a DNA strand.

Although biological engineering has existed less than a century as an academic discipline, genetic engineering is only a generation old.[19] Human proteins grown in bacterial cultures have been a commercial reality since the 1970s, when biochemist Herbert Boyer teamed up with entrepreneur Robert Swanson to found Genentech. The company started out selling cloned human insulin technology in 1982, and remains a prominent biotechnology company. The 1980, Supreme Court case *Diamond v. Chakrabarty* found that human-made organisms qualify for patents. Genetically modified organisms (GMOs) began their blaze along the bottom of the food chain soon thereafter. Since then, genetic technologies have reached far-flung industries. Scientists have grown medicine inside potatoes, cloned house cats, and grown blue roses, glow-in-the-dark tobacco, and reduced-fat ice cream made possible by protein from the ocean pout, an Arctic fish.[20] Spider genes spliced into goats cause the latter to produce stronger-than-steel, stronger-than-aramid silk, potentially for use in bullet-resistant materials. The goats would certainly enjoy that more than their experience with Kevlar fiber at the Aberdeen Proving Grounds in the early 1970s.

Splicing recombinant DNA into genomes was only a prelude. Genomes resist comprehension. They weren't designed. Half a century after cracking the genetic code, scientists have come to a predictable conclusion. Genomes are a mess by design standards. Genes overlap. They cancel each other out. Many work in tandem. The code itself is redundant, which is often a good thing in engineering. Bridges have redundant supports to keep spans aloft if primary structures break. In a genome, six different three-base sequences (codons) order up the single amino acid serine. English has one way to end a declarative sentence: the period. Messenger RNA has three "periods" that tell the ribosome to stop making a protein. And then there's "junk" DNA—the 90 percent of the genome that either doesn't do anything or doesn't do anything that scientists understand yet.

People like things neat. We like easy parts and modules that plug into each other. That reality governs the way children's toys are designed and mechanical systems like the Fords and Chevys reverse engineered by Kiichiro Toyoda. Maybe jigsaw puzzles and system design satisfy the same part of the brain, the quest for neatness, for logic, or rationality.

The trouble with our penchant for 90-degree angles and countergravity measures is that it doesn't work in the natural messiness of the biological world. Linear configurations are easy to take apart but nonlinear ones are not meant to be reverse engineered. The Model T was designed to be assembled and taken apart. The path of carbon through photosynthesis was never "meant" to be revealed, let alone mimicked, however primitively, in government labs. Proteins are too complicated to design from scratch and build. Biologists understand how DNA codes for a protein, and how RNA translates it into a protein. But predicting a protein's shape, let alone its function, from a bare string of amino acids has been a target of research for many years—and remains a massively complex problem, despite continual advances.

We can neither understand nor reproduce biology's modularity. Biology is the ultimate in modularity. Cells break down ingested proteins into amino acids, to build other proteins. Natural molecules involve intricate modularity of functional groups and triangles, additions and subtractions. All life—all matter—is built from atoms that are in their own way tiny modules, under the influence of energy flows and their own physical proclivities.

Synthetic biologists have to start small, building proof-of-concept devices that lay experimental foundations for their theories about imposing a rational paradigm on biology. A noteworthy study in 2005 showed how to rationalize the genome of the T7 bacteriophage, one of a class of viral snippets that kill bacteria. Biologists have studied T7, an E. coli–killing phage, for more than half a century. The virus is useful because its simplicity provides a good proxy for how much of genetic information is actually functional, and how much of that functionality scientists understand at this point.[21] Modeling in silico helped elucidate how the organism functions, but the real-world virus eludes complete understanding. So they made it easier to understand.[22]

Drew Endy, a BioBricks scientist and thought leader, and two colleagues reengineered and rebuilt the T7 genome, making it more modular. Endy has four amber bottles in his office, each filled with adenine, guanine, cytosine, and thymine, the nucleotide bases. "These each cost $250," he says of the materials made from sugar cane. "There's enough here to make 30 copies of the genome of every person on the planet." (That's still several order of magnitudes short the amount of DNA in a human body.)

T7 has 39,937 bp, which make up about fifty-six genes. The researchers edited out 11,515 bp and put back in 12,179—more than six hundred edits or additions. As a result, they emerged with a new organism, a neater version of T7 that works like the natural variety, with functional differences. In a nod to computer software engineering practice, they named their new organism T7.1. More ambitious projects may only be a matter of time and money, perhaps not in that order. Says Endy, "There is no technical barrier to synthesizing plants and animals. It will happen as soon as anyone pays for it."[23]

The T7 study was notable because it was early confirmation for ideas that previously had been only theoretical. But the thing about science is that some years-long research programs match ambitious or counterintuitive predicted results with fanfare, whereas many others don't. Endy acknowledges that his T7 study needn't have worked. But it did.

Synthetic biology is growing up with more self-conscious safety concerns than any field other than nuclear power. Like nuclear power, synthetic biology has peaceful and potentially bellicose uses. Unlike the origin of nuclear power, the peaceful applications drive the science. The threats from synthetic biology emanate from the dropping price of synthesizing genetic material of known pathogens and proliferation of synthesis technology. The labor-intensive nature of DNA work had served as a de facto barrier to wrongdoing, but technological change is threatening to make genetic tinkering easier. With cheaper, more widespread DNA synthesis machines, would-be terrorists can take advantage of the decoupling of genomic information, and the means to reproduce it, and ship it overnight virtually anywhere in the world. It would be easier for ill-intentioned individuals or groups to acquire dangerous cells or viruses made in basement labs.

Scientists and law-enforcement officials, across national boundaries, are trying to put in place a biosafety regime that accomplishes three ends. It must make the technology safe, without impeding scientific progress, and apply internationally.[24]

The biggest risks remain natural. There is a compelling risk that many thousands or even millions of people will die in coming years or decades if a pandemic flu akin to the 1918 virus, which killed 50 million people, struck. Should the pandemic flu threat and our lack of preparedness for it not cause concern, you might consider that the 1918 flu virus has been sequenced and synthesized from a tissue sample retrieved from a frozen victim uncovered in Alaska. Prominent technologists normally at each others' throats linked arms to condemn this bit of viral reconstructive surgery as an extremely bad idea. Bill Joy, formerly of Sun Microsystems, made headlines in April 2000 for writing an essay about the potential catastrophe of good biotechnology gone bad. In 2005, he and super-futurist Ray Kurzweil locked arms to proclaim that the resurrection of the 1918 flu virus "is extremely foolish. The genome is essentially the design of a weapon of mass destruction. No responsible scientist would advocate publishing precise designs for an atomic bomb, and in two ways revealing the sequence for the flu virus is even more dangerous."[25]

In June 2007, the Venter Institute, in Rockville, Maryland, announced that by transplanting a genome, it had turned one species of bacterium into another. *Mycoplasma genitalium,* one of the tiniest-known organisms, was sequenced in the 1990s. Craig Venter eyed it as a part of his Minimal Genome Project, a search for the smallest possible genome. *M. genitalium* has just 517 genes and fewer than 600,000 bp—a great target for investigating a biological curiosum: What is the minimal number of genes an organism needs to live? A specific estimate came in 2006, when a Venter colleague declared 386!—the number of genes on which a *M. genitalium* can survive after knocking out noncritical material.

Of all the claims of synthetic biology, few are bigger than those of Venter, who sees a world fueled in part or (probably) in whole by human-made bacteria that produce hydrogen from water and munch carbon from the air. His most dramatic success in the post–Human Genome Project era came when his institute announced it had achieved the world's first bacterial genome transplant. They purified DNA from one organism, *Mycoplasma mycoides,* and transplanted it into a related species, *Mycoplasma capricolum.* After several generations of reproduction, the latter turned into the former. Venter likened it to installing a piece of software that turns a Mac into a PC. It's more like reaching into a Volkswagen Rabbit's glove compartment, replacing the owner's manual with that of a Volkswagen Beetle, and watching the former turn into the latter. Venter told journalists an energy breakthrough could be around the corner. Several months later, Venter's team synthesized and installed a complete genome. Transformative energy technologies, however, remain distant products of this work.

Venter, Endy, and the many other synthetic-biology pioneers are simplifying biology at a time when it is becoming more complex than ever.

The Central Dogma of molecular biology becomes less and less central every year.

Francis Crick, codiscoverer of DNA's double helix, first proposed a Central Dogma for molecular biology in 1958. Another decade of worldwide research led to challenges, and he revised it in 1970. Since then, or at least until recently, RNA has largely been treated as a middleman, the product of genes that help make proteins, in accordance with this paradigm.[26]

In the last decade, RNA's role has expanded considerably, launching in part off of the 1998 discovery of RNA interference (RNAi), a double-stranded variant of RNA responsible for regulating—turning on or off—the expression of genes. More recent work has shown that swaths of DNA that do not code for proteins are nonetheless transcribed by RNA. The RNA does something with it, and molecular biologists spend a lot of time, energy, and money these days determining what these regulatory functions are. Genes have had a rough several years. The first shock came at the end of the Human Genome Project, when it became clear that the parts list for building a person is just 21,000 or so genes in size. Now it turns out that much of the human genome is transcribed for purposes other than protein making, genes' forte. The very definition of gene is changing.

How genetic information directs construction of an organism is the next great adventure in molecular biology. The ENCODE Project is a consortium of three dozen labs that in 2007 published results of intensive study on 1 percent of the human genome in action. The goal was to deepen understanding beyond a readout of the genome, into a better sense of how the genetic machinery operates. They emerged with evidence showing that much more of the genome is transcribed into RNA than previously documented. They need to make sense of why protein-coding genes, the end-all and be-all of biology for decades, make up such a small percentage of the overall genome—and what the rest of the non-gene genome might do.

This has led scientists to redefine what a gene is in the first place. The computer programming metaphor is taking on increased respect. A "subroutine" in computer science is part of a program that performs a single task, and can be called into service repeatedly, as needed. "One metaphor that is increasingly popular for describing genes is to think of them in terms of subroutines in a huge operating system (OS)," write Mark Gerstein and colleagues. "That is, insofar as the nucleotides of the genome are put together into a code that is executed through the process of transcription and translation, the genome can be thought of as an operating system for a living being. Genes are then individual subroutines in this overall system that are repetitively called in the process of transcription."[27]

In the prologue, Shirley Ann Jackson makes the point that advances in the life sciences come at their nexus with the computational and informational sciences. They are more than metaphors for each other; they are coevolving into a single *über* science.

An alternative to understanding all of this complexity is to get out of its way. In vitro evolution and directed evolution let molecules sort out for themselves which of them best addresses a problem. RNA sticks to many, many things. Jack Szostak uses this technique to explore how chemicals might behave in an evolutionary fashion.

A similar technique is being used to look for concrete answers to some of our most pressing problems—in fuels, medicine, and industry. Enzymes are proteins or complexes thereof that perform precise activities within a cell. Some of these activities, or ones like them, might be desirable in an industrial setting. Scientists looking for new biobased automotive fuels or industrial products or medicines need catalysts able to withstand conditions and perform tasks that natural proteins do not. Enzymes—proteins that catalyze reactions—don't last long, or decay under extreme conditions. Desirable new industrial enzymes might retain their biological functions, but be able to perform them under extreme conditions. Old enzymes might acquire new tricks.

The theoretical number of proteins is stupefyingly large. Life works from twenty amino acids. If a protein is, say, three hundred amino acids long, the theoretical pool of potential proteins sequences is 20^{300}, or about 2×10^{310} times the number of particles that may exist in the entire universe. That's not considering protein that use only 11 or 18 or 19 amino acids or that are 301, 401, or 2,000 amino acids long. These are ghastly numbers. They dwarf astronomical numbers. Technically, they are exponentially large. Viscerally, they feel like multiple infinities (yet as theoretical probabilities go it is still dwarfed by the theoretical limit of human gene sequences at the end of chapter two). The number of functional proteins within that pool is a vanishingly small proportion (but still quite a few!). Only the minutest fraction of potential protein sequences might be useful, so synthetic biologists begin their search for novel proteins similar to ones that they see operate in nature, ones likeliest to work.

A protein's shape determines its function. By altering that shape, slightly, and a gazillion times, and testing them for desired properties, scientists try to tease novel, mutant proteins from the heaps of theoretical nonfunctioning garbage. Most of that infinity of possible proteins are just amino acid strings that wouldn't even fold, let alone exhibit useful properties. Frances Arnold is a bio- and chemical engineer at Caltech. The absurd number of theoretical proteins led Arnold to invent ways to direct the evolution proteins. Her interest grew from directed-evolution studies on RNA in the early 1990s, particularly Gerry Joyce's work at the Scripps Institution, which involved directed evolution of nucleic acids.

Arnold looked for novel, functional proteins lurking one to three amino acid changes away from natural ones. It's easier said than done. Changing one amino acid in a protein 300 amino acids long yields 5,744 possible variants. Two amino acid substitutions leaves 16 million new possibilities, and three—30 billion.[28]

It's hard to think of a more emblematic union of the blind evolutionary drive to mutate and adapt with the human technological drive to find or make useful things than directed evolution. For one, it *is* evolution (with a human nudge). The principles of evolution apply to DNA or RNA snippets, or proteins, as aptly as to species. Still, experimenters must impose their own goals on the process if they want a useful outcome. Normally, evolution lunges blindly in whichever direction conditions don't discourage. Technology fixes itself on a specific goal, tries a few solutions, and ends up with one. That's how we end up with fleets of passenger cars that all have four wheels and the driver on the left or Microsoft Windows or the Boeing 737 or the qwerty keyboard. "All human inventions are evolutionary," Arnold says. "We accrete the good things and throw away the bad ones." Using directed evolution, biochemists fix a goal and simultaneously flatten more than 10^6 paths to reach it.

The difference that directed evolution offers over traditional synthetic chemistry or protein-structure studies is simple. The first way it differs is in the ability to blend the scientist's goal-oriented quest with the parallel processing power of nature. Instead of studying one molecule at a time, they can study millions and billions. The second is that scientists can look for novel functions without bothering to learn the shape of the molecules themselves. They just build variants of known molecules, reproduce them in great numbers, and test for desired properties, agnostic of structure.

The medical and energy industries are eagerly tapping their feet for the fruits of this creativity. Arnold has started two companies, hoping to make nature's robots go to work for the private sector. One of the proteins she has worked on is cytochrome p450, a large class of enzymes nearly universal to living things. They help build natural steroids, ready arsenals deployed in "interspecies chemical warfare,"[29] partially detoxify harmful compounds—including cigarette smoke—metabolize barbiturates and other drugs, and break down unfamiliar foods.

The designation p450 indicates that the molecule absorbs light at a wavelength of 450 nm, squarely in the indigo blue part of the light spectrum, making the resulting protein red as blood.[30] P450s are extremely good at popping oxygen atoms into molecules. Some specialize in CH bonds, a useful catalyst to have, if you want to turn gaseous methane (CH_4) and ethane (C_2H_6) into liquid fuels methanol and ethanol. Arnold has built a new library of p450s by combining snippets of natural proteins and then using directed evolution to optimize the desired properties. Directed by computer-modeling results, the lab assembled chimera p450s and used evolutionary search tools to identify useful molecules, amplify them, and conduct further research. The potential benefits are huge. Some of the p450s might help build drug metabolites—enzymes that help the body break down therapeutic chemicals. Others might prevent pesticides from harming beneficial microbes in soil.

The home run may be in fuels, where Arnold is finding ways to convert cellulose efficiently into 4-carbon butanol. Her start-up fuels company, Gevo, plans to commercialize its novel way to make butanol into a transportation fuel. In this space Gevo has company from high-profile start-ups LS9, which is trying to bring to scale gas-tank ready fuel grown in bacteria; Amyris Biotechnologies, which is working on gasoline and diesel substitutes; and Craig Venter's Synthetic Genomics. These technologies are still young, expensive, and relative to traditional fuels, inefficient. But the invisible hand of economics may be moving in to fix a fuels problem it created in the first place.

Human methods are laughably simplistic compared with the systems that nature has evolved over billions of years. We sometimes produce more economic and elegant designs than the redundant, chaotic designs produced by trial and error evolution, but nature's finest creations, which is to say any of them, are too complex for us to fully comprehend, much less duplicate. Biotech will have useful applications, but we are not going to be able to replicate nature's services, and so we had better protect those services.

And if we ever do learn to live within the short-term carbon cycle, advocates wait in the wings to take it to the next step: pack up the biosphere and dig in to foreign worlds.

12 | The Adventures Ahead
Life With Carbon,
Civilization Without?

> riverrun, past Eve and Adam's, from swerve of shore to
> bend of bay, brings us by a commodius vicus of recircula-
> tion back to Howth Castle and Environs.
>
> —James Joyce

Life in general and humans in particular have abundant energy choices. Photosynthetic cells weld CO_2 into carbohydrates with sunlight. Others tap subterranean heat. If there's an energy source on Earth, cells find it. In 2006, scientists reported a discovery of microbes that live two miles underground, in a South African gold mine, living off sulfur laid down during the Precambrian and hydrogen split from water by a local uranium patch. For millions of years they have lived off energy and fuel separate from all other ecosystems.[1] Microbes turn up in virtually every crevice we peer into. They inhabit so many bizarre (to us) environments it would seem natural for them to conclude, if they only had a brain, that they are the truest Earthlings and we are the extremophiles.[2]

Generally speaking, "carbon," the buzz word, is confined to meanings in either climate-change discussions or high-end sporting goods. When I began researching *The Carbon Age,* the war in Iraq was a mess at six months old, the Atkins "no-carb" diet was careening toward its spectacular blowout, and the private sector was gradually embracing the implications of global warming. All these news items were fundamentally—chemically—about hydrocarbons, carbohydrates, and the result of their combustion—carbon dioxide. When you drive your car carbon dioxide flies out the tailpipe. When you run, carbon dioxide flies out your windpipe. The world of carbon lay beneath these stories, waiting to be found.

A key characteristic of carbon compounds, one that's not true of most inorganic ones, is that they burn.[3] Something unusual happened when human ancestors lit their fire. They opened a new source of biological energy on the planet—as if there weren't enough already—energy stored in carbon molecules but consumed and controlled *outside* cells.

There are no walls between hydrocarbons, carbohydrates, and carbon dioxide, other than the ones that nineteenth-century European chemists invented to organize their observations. The Earth doesn't store petroleum by the barrel. Nature does not categorize its creatures into two-word Latin species. The atmosphere doesn't "think" about whether it is putting on too many carbs. Carbon, hydrogen, oxygen, nitrogen, phosphorus, and sulfur pour through the Earth system, driven by fluid dynamics, the crawl of the continents, and the Sun's heat.

If there is a vein of hope in the accelerating juggernaut of fossil fuel burning, it has to do with the quality of fuels used over time. Nations, industries, and individuals are talking about stripping carbon from our industrial energy supply, hopefully in time to make a difference. Sunlight, wind, water, geothermal energy, and bioengineered fuels from bioengineered bugs will provide increasing amounts of power in the future, as fossil fuels become expensive, both through scarcity and the price of emitting CO_2. To appreciate this trend, one needs to swallow hard and overlook for a moment the quantity of fuel we burn to look at its quality.

Much has changed in a short time, and it will only speed up. At this writing, coal-burning generators are beginning to earn ire once reserved for nuclear plants. Nuclear plants provoke fears of meltdown; yet coal burning is already fueling a very literal global meltdown. Unfortunately, we are wedded to coal use for the foreseeable future, unless industrialized nations agree to forego electricity and heavy manufactured goods. Carbon capture and storage technology would keep the coal industry in business into perpetuity, enabling it to catch CO_2 and pump it back where we found it. "The carbon belongs underground," says Susan Hovorka, a geological engineer at the University of Texas. "I say, put it back." Movement is afoot, including Hovorka's work, a demonstration project that injects CO_2 1,500 meters underground in aquifers. The technological know-how is often the easiest part. Making it economical, safe, and politically palatable takes time—time that we have less and less of.

Carbon capture and storage is perhaps the most important of all the new energy technologies talked about today. It would allow industry to keep burning coal—which it shows no sign of stopping anyway—but dispose of its carbon. Disposing of coal's waste gas would also encourage a long-term trend. The carbon-hydrogen ratios in fuel have changed over time, shifting from the former to the latter. Wood and charcoal fueled life and invention more or less until the nineteenth century. Most of wood is cellulose, a carbohydrate. It burns easily, heat carrying away the hydrate (H_2O) and leaving just the carbon, as charcoal.[4] Lignin, the other major component of wood, is a carbon-rich affair, a network of carbon rings, and branching chains. A log fire oxidizes about ten carbon atoms for every hydrogen atom. That's why they leave so much ash behind. The carbon concentration overwhelms oxygen's ability to catch up. The proliferation of railroads promoted coal, which is clean only by comparison to wood. Coal oxidation ideally produces CO_2 over H_2O in about a 2:1 ratio. Automobiles made petroleum the fuel of choice. Its hydrogen outnumbers carbon 2:1. Natural gas became a major fuel for the generation of electricity in the 1990s, prompting construction of gas-burning power plants across the United States. The main component of natural gas is methane, the most energetic hydrocarbon, offering to oxygen four hydrogen atoms for every carbon.

Decarbonization is about as clumsy sounding as words come, but it's expressive. Jesse Ausubel of Rockefeller University began using the word in 1991 to describe a trend discovered in the 1980s: the carbon intensity of energy was in historical decline, in parallel to the acceleration of fossil fuel burning. Ninety percent of economic activity was fueled by carbon, via wood and coal in 1800. The hydrogen content of fuels caught up by 1935 or so, as developed nations switched to oil and gasoline stations popped up on street corners. It may be too soon to tell whether the rise of China as a major energy consumer and a resurgence of coal in the United States spoil this trend.[5]

Hydrogen combustion packs more of a wallop than carbon. It's good to keep in mind Peter Atkins's extolling of carbon's mediocrity. H-H bonds release 482 kJ per mole when it reacts with oxygen to make water. C-C bonds release 252 kJ per mole on the road to carbon dioxide. Bodies and machines burn carbon because it is ubiquitous, the Velcro of life, and a good fuel—not a great fuel.

The energy trend toward hydrogen combustion and away from carbon is threatened, at least in the short term, by the resurgence of coal as a power source, mostly in the world's top two CO_2-emitting nations, China and the United States. There's a lot of coal in the world and at this writing, very few people are capable of just saying no. Coal is easy and cheap. Coal industry officials have called the United States the "Saudi Arabia of coal" at least since the 1973 oil crisis. Indeed, there is more sunlight stored as coal beneath the United States than there is oil beneath Saudi Arabia. Natural gas combustion would extend the decarbonization trend, but it is difficult to transport and has proven susceptible to wild price swings, as the electricity sector grew an affection for it in the 1990s. It must be frozen into a liquid at −160°C (−256°F), and then thawed, *very* carefully, once it reaches port. For the moment, the United States receives about 85 percent of its natural gas from Canada.[6]

The industrial energy supply is poised for such crisis that some forward thinkers are looking into space as the ultimate carbonless energy source. Solar power might be harnessed without relying on the Sun. Scientists mastered uncontrolled nuclear fusion more than half a century ago, when the United States tested "Mike," the first hydrogen bomb. Still a science project—in some ways still a thought experiment—controlled nuclear fusion carries hope in some circles.

Helium is the second most abundant element in the Universe, after hydrogen. The Moon has approximately 500 million tons sequestered in its topsoil. Helium's atomic mass is 4, for its two protons and two neutrons. When stars 20 percent larger than our Sun reach 100 million degrees or so, helium nuclei fuse into beryllium-8 frequently enough that a third helium can join in and create carbon-12. That's the triple-alpha process, and the way carbon is born into the Universe.

Fusion reactors have operated experimentally for more than a decade. In the 2004 movie *Spiderman II*, the twisted genius Doc Oc, creates a fictionalized tritium-fueled fusion reactor that nearly destroys the city. The screenwriters did some homework before taking their flights of fancy. Tritium is a hydrogen isotope of atomic mass 3. Its nucleus contains one proton and two neutrons. The fusion of tritium with deuterium, the hydrogen isotope of mass 2, yields helium-4 and neutron radiation. It's too much radiation for safe energy production at scale. Six nations and the European Union have agreed to construct a tritium-deuterium nuclear fusion reactor in Cadarache, France, by 2015. The United States is a committed provider of talent, materials, and funds, about 9 percent of the total, or $1.122 billion.

A handful of labs around the world look beyond even the carbon-free tritium fusion reactor, to its even further-off sequel. Theoretically, helium-3 fusion would emit less radioactivity than tritium reactors. On the other hand, the reaction doesn't proceed as easily. Two logistical problems confront helium-3 reactors. First, other than an intriguing experimental reactor at the University of Wisconsin–Madison and much hullabaloo about Chinese energy goals, the technology does not exist. It's not even on the horizon, and not a target of major federal funding. Second, the capital of helium-3 is even farther away than Saudi Arabia—the Moon.

The line between science and science fiction keeps moving, and might currently run through the future of helium-3 research, a far-off hope, eternally fifty years away. The solar wind blows helium atoms into the solar system. En route, cosmic rays knock loose neutrons, turning helium-4 into lighter isotopes. They hit the moon and stay there, gathering in abundances of 13 ppb (parts per billion), higher in some locations. The Earth's atmosphere prevents the accumulation of the isotope here. Apollo 17 astronaut and geochemist Harrison Schmitt calculates that at $40,000 per ounce, 220 pounds of helium-3 laboriously baked from 5.2 million cubic yards of soil might power a medium-sized U.S. city for a year.[7] The material would fuel so-called second-generation nuclear fusion reactors, the power of the Sun in a can on Earth. The United States in 2005 used the energy equivalent to 40 tons of helium-3.

Helium-3 fusion theoretically might work for the same reason George Gamow's initial version of the big bang hypothesis failed. George Gamow thought the big bang created all the elements by fusing hydrogen to every higher mass. He ran into a wall at mass 5. Helium-3 or tritium combine with deuterium, trying to make a mass-5 nucleus, only to immediately decay to helium-4 and a proton or neutron, and energy. That's why the triple-alpha process is so important. Helium nuclei leap over the smallest atomic masses to carbon-12. If helium-3 reactors ever work, humans would be taking advantage of the same failed nuclear reaction that requires stars to make carbon the way they do.[8]

Humans looked skyward long before we landed on the Moon. As early as 1869 the American writer Edward Everett Hale envisioned human colonization of space in his novel, *Brick Moon*. A large, brick sphere designed to hang in the sky as a maritime beacon accidentally launches with construction workers still inside it. Fortunately, they have ample provisions and decide to live in space, sending Morse code signals to Earth and shuttling back and forth as necessary.[9] Space colonization dominates science fiction. The original *Star Trek* TV show—launched five years after Russian Yuri Gagarin became the first human in orbit—had a famous "five-year mission . . . to boldly go where no man had gone before." V'Ger, the mysterious enemy in the first *Star Trek* feature-length movie famously thought of human beings as inferior "carbon units." The notion of a "galactic empire" first appeared in Isaac Asimov's Foundation novels and was later immortalized by George Lucas in *Star Wars*.

Scientists' theoretical musings on the possibilities of space life trail novelists by several decades. Vladimir Vernadsky wrote in his 1944 essay about the Noösphere, "Fairy tale dreams appear possible in the future; man is striving to emerge beyond the boundaries of his planet into cosmic space. And he probably will do so."[10]

In 1960, the eclectic physicist Freeman Dyson published a thought experiment in the pages of the journal *Science*, where he calculated that it is theoretically possible to reconfigure the entire mass of Jupiter into a 2- or 3-meter-thick revolving shell twice as far from the Sun as Earth. Humans could slowly inhabit the shell, culling all the energy the Sun puts out, rather than the paltry beams the Earth absorbs.[11] A civilization that advanced would presumably find something out there to breathe.

In the half century since then, space enthusiasts have held forth on space colonization, from the administrator of the National Aeronautics and Space Administration to Stephen Hawking to Newt Gingrich. NASA and nonprofit groups hold conferences on the engineering, logistical, political, economic, and sociological aspects of space settlements. Many scientists and nonscientists speak of space colonization as inevitable, either for chasing down the "final frontier" or because a catastrophe makes Earth inhospitable.

At least one visionary, and his backers, is reinvigorating aerospace goals more modest than Mars settlements. Many scientists have conducted extraordinary work laying groundwork for a future in space. Burt Rutan is founder of Scaled Composites, LLC, a Mojave, California, firm that designs and builds advanced aerospace vehicles made largely of strong, lightweight carbon composites. He has gained public prominence in recent years, inspiring people to think about space flight as no one and nothing has in more than a generation. His SpaceShipOne flew twice into space within fourteen days, capturing the $25 million Ansari X Prize. His Voyager aircraft flew around the world nonstop in nine days. Space-ShipTwo is expected to take well-off space tourists into gravity-free space within the next few years. We might never get to space permanently, but the drive to do so will likely lead to many useful things. Rutan has said, "We're entering a second generation of no progress in terms of human flight in space. In fact, we've regressed. We stand a very big chance of losing our ability to inspire our youth . . . I feel very strongly that it's not good enough for us to have generations of kids that think it's okay to look forward to a better version of a cell phone with a video in it. They need to look forward to exploration. They need to look forward to colonization. They need to look forward to breakthroughs. We need to inspire them because they need to lead us and help us survive in the future."[12]

The current lack of any plan, real or imagined, to permanently occupy space should not be confused with its impossibility. That it is preposterous to common sense, current federal budgets, and public priorities should not necessarily close minds. The history of technology both warns against underestimating ideas and human potential for firing them into reality, and misfiring them into oblivion. Many things we now take for granted were once considered patently absurd. A century and a quarter ago there were no cars. Today there are 800 million in service. The exponential rate of growth in computer storage is a perfect example of how the future sneaks up into the present. Space is a dangerous business. Space settlement seems prima facie absurd for many reasons. At this writing, it's difficult enough to make New Orleans or Baghdad safe, and both those cities share an atmosphere with oxygen, radiation-blocking ozone, and a powerful carbon greenhouse. Future space voyagers are likened to Christopher Columbus and the great European explorers. The latter had the benefit of running aground in the same biosphere, with plenty to eat. Life drives and is driven by the geochemical cycles of nature. Leaving it without a fail-safe, portable climate would be suicide.

Even if we decarbonize our machines, build helium-3 reactors, or colonize space, mitochondria will always prefer carbon. Perhaps we can feed them on the Moon. The late Larry Haskin, an American geochemist, observed in a thought experiment that atoms sufficient to make cheese are distributed throughout the lunar topsoil. Pale gray Moon dust sequesters the right elements to assemble food of any variety and presumably in any color. No one has tried to age a Parmesan wheel from dust, but at least the atomic ingredients are there.

Cheese could be just the hors d'oeuvres. Atoms of CHONPS elements are so abundant in lunar soil, a cubic meter might provide enough for "the chemical equivalent of lunch for two—two large cheese sandwiches, two 12 oz sodas (sweetened with sugar), and two plums, with substantial N and C left over."[13]

Oxygen makes up about 44 percent of Moon rocks. Hydrogen is extremely rare within the Moon itself, but the solar wind sprinkles the surface with ions that over 4.5 billion years have left about 100 g of hydrogen per cubic meter. Nitrogen exists in about the same abundance. Chemists who examined Apollo-era samples found about 1.8 kg of sulfur, 1 k of phosphorous, and dashes of noble gases. The Moon's soil contains nearly 200 g of carbon per cubic meter, about 37 percent of the earthly equivalent. "In terms of life support these amounts are enormous," wrote Haskin. "Collection of even a small fraction of the Moon's budget of H, C, N, P, S, and other elements essential to life into a suitable environment on the Moon would support a substantial biosphere."[14] The key words here are "suitable environment." They are undefined and obscure the most complex engineering problem ever devised—how to replicate Earth's geochemical cycles in an airless, lifeless crater.

"Suitable environment" hangs over Haskin's thought experiment, indeed over much of space planning, as a monumental "TBD" (to be determined). Heating the atoms from soil shouldn't be too difficult. That's how oil is refined and ground beef is browned in a skillet. Moon visitors (or residents) could bake the atoms out of the soil. Hydrogen would dislodge at about 700°C, energy that would have to come from somewhere. Moon visitors could pump waste heat back into the reactor to save energy. The gases would come out as water, carbon dioxide, ammonia, and the poisons carbon monoxide, hydrogen cyanide, and sulfur dioxide (think acid rain). After further combustion, the products could be separated by supercooling. Moon colonists could theoretically assemble or grow them into two cheese sandwiches, plums, and soda. In current dollars, it would probably be cheaper to invent and fly space modules from the Moon to a convenience store near Houston's Johnson Space Center, an inconvenient 250,000 miles away. Humans have traveled round-trip to the Moon, but no one has ever synthesized lunch from dust. It took Robert Burns Woodward and his seventeen assistants four years to make tiny amounts of chlorophyll from scratch, in an oxygen-filled atmosphere, with plenty around to eat.

Space travelers would need more than lunch. They would need a portable biosphere to convert the waste remains into plant food, so that the plants could resupply human nutrients. As the eclectic scholar Vaclav Smil has put it, "Advanced civilization, humane and equitable, could do very well without Microsoft and Wal-Mart or without titanium and polyethylene—but, to choose just one of many obvious examples, not without cellulose-decomposing bacteria."[15] A permanent space settlement would need a biosphere with it. The problem: The only one we know of needed 4.5 billion years to enable conditions safe for modern humans.

Another logical problem posed by space settlement is transporting enough people into space. How many people can fit in a tin can launched into orbit? This philosophical question led—through metaphor, and the strength of such personalities as futurist R. Buckminster Fuller—into the much bigger problem of how many people fit on Earth without ruining the planet. In the wake of Rachel Carson's 1962 book *Silent Spring*, which triggered the environmental movement, ecological questions became real. We really might ruin this planet. Perhaps we might find refuge on another world. "Terraforming" means the relandscaping of sterile foreign worlds into Earth-like verdant wilds. The astroecologists' goal was "to combine natural components with mechanical shortcuts," wrote Eugene Odum, who with his brother Howard layered engineering principles into ecology.[16]

By the late 1950s and 1960s, Buckminster Fuller had already long established his vision for simpler, smarter geometry and engineering for human systems. His Air-Ocean World Map, first sketched out in 1927, was only an early foray into his polyhedral world that included inventing

*R. Buckminster Fuller's 250-foot-wide structure housed the U.S. pavilion at
the 1967 Montreal World's Fair. Fuller's ideas influenced the development
of thought on humans' place within Earth's ecology.*

geodesic domes. Fuller saw the ecological lessons in space research earlier
than anybody else. Fuller's familiarity with military, specifically naval research into the "closed chemical circuit of ecologic . . . living of moon-
rounding men," led to his application of these principles to the planet
itself. "We are all astronauts," he answered those who would wonder
about the experience of manned spaceflight.[17]

Princeton's Gerard K. O'Neill pushed the ball forward in the 1970s, ar-
guing that "careful engineering and cost analysis shows we can build
pleasant, self-sufficient dwelling places in space within the next two decades, solving many of Earth's problems."[18] Over the next decade, Buckminster Fuller's intellectual offspring helped establish an ambitious
experiment to show that a mini-Earth, Biosphere 2, was a workable concept, with enough will and capital. Phil Hawes, the architect of Biosphere
2's futuristic-looking geodesic domes and oblique greenhouses, is a disci-
ple of Fuller.

The founders hoped to show that living things could continue to coevolve and biogeochemical cycles of carbon, nitrogen, sulfur, phosphorous, and the other elements of life could perpetuate themselves in an ecosystem severed from Mother Earth—Noah's Ark, where no organism was restricted to a single male and a single female (particularly helpful among species that reproduce themselves asexually or by other means that don't require individuals to be in the same room). Synthetic mobility, synthetic chemistry, and synthetic biology have their logical end in synthetic ecosystems.

The design and engineering of a closed-loop or portable ecosystem is as daunting a task as scientists have ever faced. The Earth took 4.5 billion years to sprout humans. Packaging the biogeochemical cycles of nature and spinning them into a gravityless vacuum might not be so easy. NASA and its international counterparts have launched people and cabin-sized atmospheres into space since 1961. These environments tend to foul rather quickly. Human waste is bagged and disinfected, ejected into orbit or freeze-dried for the trip home. No mission has broken down waste into chemical components and reassembled it into cheese sandwiches, plums, and soda. Scientists study plant and microbial species trying to find hardy organisms that could turn human wastes—carbon dioxide and postdigestive matter—into plant food.

A quick look at soil research shows how little science has penetrated the complex web of even tiny ecosystems. Microbes keep the carbon cycle spinning by moving nutrients through living and once-living matter, soils, rock, and water. They clean water, break down toxins, and turn waste back into food. Yet scientists have looked at fewer than five thousand distinct species of bacteria and archaea, perhaps 0.1 percent of their total.[19] Most would die before they can be taken to a lab to study. The silent spins and ellipses of heavenly bodies are elegant toys compared with the hysterically teeming complexity of microbes and their chemical footprints in a square inch of soil. "Shotgun sequencing," of the type Craig Venter has applied to microbes in the open sea, has made its way into the dirt, which stores double the carbon of the plants it bears and the animals that eat them.

A viable synthetic ecosystem would have to replicate the Earth system from atmospheric mix down to the interactions of microbes that no one has ever seen. Vaclav Smil, shooting from the hip, said of the space-colonization movement in general, but no one in particular, "They are idiots. They don't understand biology."[20]

Somehow, a defiled Earth with rain forests at the poles still seems homier than a cold atmosphereless desert, such as the Moon or Mars, or any other planet too far away to reach in a lifetime. Besides, as Princeton's J. Richard Gott III has argued, we may not colonize space not because we are unable to, but because species simply rarely live up to their own potential.[21]

Biosphere 2 represents by far the most aggressive and interesting eco-engineering experiment. Usually, accounts of Biosphere 2 point out failures in execution, tragic flaws that circumscribed Olympian ambition within human limitations. If the goal was to make a mini-Earth, as livable and sustainable as the planet itself, the project failed at conception. Biosphere 2 rose from the desert in Oracle, Arizona, in the mid-1980s, a futuristic landscape of glass and white. The greenhouse enclosure itself covers 3.15 acres and houses seven distinct ecosystems. Many accounts of Biosphere 2 open with its cost—$200 million. That's less compelling a complaint in the age of $200 million Hollywood movies, although the movies generally make most of their money back. As far as money-losing propositions go, biospheres seem a nobler long-term investment than poorly attended blockbuster sequels, particularly if the project needs no public funding.[22]

The energy needed to run biosphere's energy needs was transmitted in from the outside, and not just as sunlight. The team went to great lengths to make the place airtight. The complex rests on a massive stainless steel plate that separates it from Earth. But energy—the alpha and omega of the biosphere—was taken for granted. Transmission lines delivered electricity to the Biosphere 2 from a nearby natural gas–fired power plant. The palace was designed to be a prototype home for humans on the Moon or Mars. But there are no gas-powered electricity plants waiting for us to plug in our biospheres on other planets. Biosphere 2 aimed to confront the most challenging engineering problems imagined—except for the most important one: energy.

More than 3,800 species of plants, animals, algae, and microbes filed in. More than three quarters of all vertebrates went extinct. Most insects went extinct, including all of the pollinators. Deserts became grasslands. As if they knew their own symbolic cache, crazy ants (*Paratrechina longicornicus*) took over habitats, with cockroaches and katydids. The morning glory vine inched throughout the glass world. Trees weakened and collapsed.

The goal was to set up an Eden in the desert and appoint four Adams and four Eves to tend it, sealed from the outside world for two years. That task proved impossible. Taking parts of the Earth and putting them under one roof didn't make a mini-Earth, growing in tandem with Earth itself. It taught a lesson in managing a complex web of life, that lesson being: it's way too hard. The first team to move in and seal themselves off from society—sort of—started in September 1991.

Nearly a year and a half into the venture, a dangerous mystery emerged. The oxygen concentration of the complex had gradually fallen from 21 percent—its abundance in the air—to 14 percent, the equivalent of living atop a 17,500-foot mountain. The biospherians, piped in 23 tons of molecular oxygen to replenish normal levels, which began their slow retreat again.

The disappearance of oxygen puzzled biospherians and outside experts. If oxygen was declining, common sense would predict a rise in its complement, carbon dioxide. That's how Biosphere I (Earth) generally works. The scientists found no such correspondence. Oxygen and carbon dioxide were both disappearing, which didn't make sense.

Before the experiment, the University of Arizona's Environmental Research Laboratory had run computer models to determine the optimal percentage of organic content in food-growing soils. They came to the conclusion that soils should contain about 4 or 5 percent organic material. The actual amount brought in was closer to 30 percent. As a result, bacteria that live in soil munched their bounty into carbon dioxide and water breathed in lots of oxygen. That solved one puzzle.

The carbon dioxide was a different matter still, one with symbolic implications for humanity's relationship toward the Earth in general. After the eight scientists of Mission One left Biosphere 2 in September 1993, Biosphere 2's owners asked scientists from Columbia University's Lamont-Doherty Earth Observatory to perform a postmortem on the missing oxygen and carbon dioxide. Jeff Severinghaus, then a Columbia graduate student, discovered after a tip from his scientist-father that the structure of Biosphere 2 itself was absorbing the carbon. The concrete structure contained calcium hydroxide. The carbon dioxide wafted up from the soils and reacted with it to create calcium carbonate.[23] Biosphere

2 was eating itself. In the end, the biosphere failed because the biospherians forgot to take themselves, and the effects of carbon dioxide, into account.

As Biosphere 2 rose from the desert in the mid-1980s, a more indirect legacy of Buckminster Fuller materialized 1,000 miles to the east. A simple description of the event loosely mirrors Fred Hoyle's prediction of the previously unknown energy level in the carbon-12 nucleus. A visiting British scientist has brought his idea for an experiment to an advanced American laboratory; it takes ten days or so, and the results are astonishing.[24]

On September 1, 1985, Rice University chemists Richard Smalley, Robert Curl, and three graduate students assembled to run an experiment for visiting professor Harry Kroto of the University of Sheffield. Kroto had waited a year and a half for the chance. He wanted to see if long carbon chains, cyanopolyynes, might form under conditions similar to a star's gaseous envelope. From there, Kroto argued, they blow into dense interstellar clouds, where they had been observed. To perform the experiment, Kroto requested time on the lab's laser vaporization cluster beam apparatus, a powerful tool invented and built by Smalley and the members of his lab, who used the room-size apparatus to study how atoms dance when you set their floor ablaze.

Laser pulses blasted gazillions of carbon atoms from a graphite disk inside the apparatus. They cooled in a gust of helium. A vacuum sucked them into another chamber, where they cooled further. The Rice researchers detected the molecules Kroto predicted they would. Problem solved. The cyanopolyynes, in a sense, had "fallen to Earth," a metaphorical phrase Kroto would invoke in subsequent years to describe what happened next, with a different molecule altogether.

That's not all they saw. Molecules containing exactly sixty carbon atoms showed up with surprising frequency. "It gradually became clear that something quite extraordinary was taking place," Kroto wrote. "Sometimes it was completely off-scale; at other times it was quite unassuming."[25]

Atoms can't count. They shouldn't know to stop working after certain magic numbers. Sixty was only the most curious anomaly they found, an aberration similar to that detected recently by Exxon researchers. Atoms usually glom onto each other haphazardly, like capsized passengers groping for a raft. Each finds a place in the molecule and holds on. Carbon molecules look like chicken wire, centipedes, heaps, branches, antlers, or twisted blobs; but this sixty-atom cluster appeared too regularly to be a twisted blob. Unaware of C_{60}'s shape, the team picked up on Kroto's use of the word "wadge." Smalley called the molecule the "mother wadge." Kroto promoted it to "god wadge."

Days passed. Researchers reset the apparatus and reran the experiment many times, curious to see how different conditions affected the abundance of this mystery molecule, C_{60}. Kroto wondered if it might resemble a cardboard astral globe he had at home and mentioned it to the others. He stopped short of calling his wife in the middle of the night, UK time. The wadge dominated conversation that night of September 9, at the group's favorite Mexican restaurant. Everyone agreed C_{60} must be some kind of closed structure.

Working at home the night of September 9, Smalley sought a low-tech solution in the kitchen. He tried to solve the mystery using hexagons cut from paper fit together like floor tiles. Interlocking hexagons was not the answer. There wouldn't be any physical reason for a carbon sheet— graphene—to stop growing at sixty atoms. Smalley taped five hexagons around the edges of a pentagon and to each other. The structure curled into a bowl. He added another pentagon and a few more hexagons. When he finished, he held a paper sphere of twelve pentagons and twenty hexagons. The next day, Smalley called a mathematician, who relayed back that the shape is technically called a truncated icosahedron: "Tell Rick it's a soccer ball."

The Rice team would take nearly a year to realize that Japanese chemist Eiji Osawa had predicted C_{60}'s existence and properties in 1970, inspired by watching his son playing soccer.[26]

Smalley dropped the model on his office coffee table the next morning, his colleagues looking on. The chemistry worked. Curl called Kroto, who was readying to leave, but postponed his trip a day. They named the molecule *buckminsterfullerene*, for its resemblance to Fuller's geodesic domes. The original paper sphere, one of the most storied artifacts in modern science, sat on Smalley's shelves for many years, until books prone to fall squashed it for the last time. He threw it out.

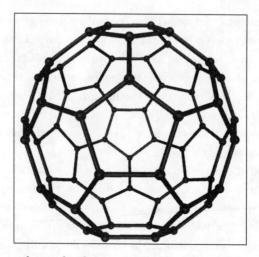

60-atom spheres of carbon were named buckminsterfullerenes for their similarity to geodesic domes.

A vial of buckyballs looks like soot. It is soot. Candles make trace amounts of C_{60}. But its remarkable, regular structure also makes it a dark crystal. "You put the last atom in, and there's a sense of completeness," Smalley said. "And that molecule—for gazillions of collisions with other 'nanothings'—keeps its identity. And it's a different identity than if it had one more or less atom. Now *that* was a revelation! *That's* what it means to be a molecule!"[27]

The molecule both incited controversy and achieved iconic status when the journal *Nature* broadcast the team's discovery in November 1985. The cover of the November 14 issue juxtaposed the molecular diagram and its larger twin, the soccer ball. It took several years for scientists to cook C_{60} in bulk quantity and confirm the work done in Houston by the Rice scientists and Kroto.

Symmetrical molecules hold special interest among both chemists and nonscientists, Kekulé's benzene insight being perhaps the most storied molecule previously. Tetrahedrane is largely a theoretical molecule, four carbon atoms arranged as the points of a four-sided pyramid. Scientists have built related molecules, under special conditions, which have side groups instead of hydrogen atoms bonded to the carbons. Dodecahedrane has twenty carbon atoms and twelve faces.[28] Cubane (C_8H_8)—a molecular cube with carbon atoms at the eight vertices—was first synthesized in the 1960s, an ornery little molecule. Carbon likes to bond at 109.5-degree angles to other carbon atoms. Shoving them into 90-degree angles makes them extremely uptight. In 2000, the University of Chicago chemist who first made cubane synthesized gram amounts of octanitrocubane, a carbon cube with high-energy nitrogen-oxygen groups attached to the vertices. Octanitrocubane might be the most powerful non-nuclear explosive theoretically possible.[29] Kroto's interest in Smalley's lab was first piqued on a previous visit, when the lab successfully made triangular molecules of silicon and two carbon.

C_{60} has brought more inspirational, or symbolic, than practical or economic value to the field. The discovery of C_{60} became a founding moment for nanotechnology—the science and applications of goods whose chief operating parts fall on the nanoscale, or billionths of a meter. "Bucky is still in school," Smalley used to say. "He hasn't got a job."[30]

Buckyballs have cylindrical cousins, called carbon nanotubes, which were discovered in 1991. They essentially "look like" rolled chicken wire, sheets of six-carbon rings curved into tubes by five-carbon rings. Nanotubes have extraordinary properties, and are used in experimental transistors, soft body armor, and electrical wire. Their remarkable resilience and ability to handle wear and tear recommend them as a future structural material for artificial muscles. Smalley spent the last years of his life engaged in research that advanced the manufacture of single-walled carbon nanotubes. He died of cancer in October 2005 at the age of sixty-two.

When I met with Smalley, in May 2004, he and I didn't talk much about nanotubes or even C_{60} that day. We talked about the carbon atom and the physical properties that make it citizen king of the periodic table. We talked about the pull of nuclei on electrons, and the pull of gifted young minds away from scientific fields. A graphic on his door showed the increase in Chinese science and engineering Ph.D.s versus those in the United States and Europe.

The third floor of Rice University's Space Science Building is adorned with frame after frame of photo, news article, award, certificate, or other memento from Smalley's career as codiscoverer of C_{60}. He'd conference with members of his labs in rooms lit only by slide projections, evaluating new methods to synthesize nanotubes in desired lengths and widths. The heart of the lab is the descendant of the cluster beam apparatus.

In Houston, the U.S. energy capital, Smalley spent his own energy bringing a message to influential conduits between him and the president of the United States, an oil-industry alumnus. He tried to communicate to the administration through these back channels how beautiful and heroic it would be if former oilmen took significant steps to curb global warming emissions. "It would be bigger than Nixon going to China," he said.

I asked, if global warming was so vital, why not cut out the back-channel middlemen and campaign publicly for change? His answer was that independence is a treasured thing, and hard earned. Independence bestows credibility on professionals of any stripe—scientists, doctors, Wall Street analysts, accountants, journalists. To Smalley, sacrificing that would amount to sacrificing the power and efficacy of his word. He visited Washington with some regularity, speaking at conferences or meeting with public officials in the hope of nudging along action on the climate and energy crises. "Action on global warming can be driven by heroic leadership or by events," he said. His eyes darted away and then back. "It'll probably be events."

ACKNOWLEDGMENTS

Every author owes a debt of gratitude to supportive individuals, in the way that someone stranded in the desert for three days owes a debt of gratitude to an oasis of banana trees and cool, fresh water. Before I thank people who supported me during research and writing, I would like to acknowledge the research associates I spent the most time with.

This book would have been inconceivable without electronic research databases, particularly Google Scholar, ISI Web of Science, ProQuest, Columbia University's on-campus resources, and the online archives of many peer-reviewed scientific journals. Google Scholar stands out for the nimbleness with which it tracked down journal articles that eluded even the Library of Congress intranet. At no previous point in history—before Internet search through professional literature—would it have been possible for a single individual without a research staff to, if nothing else, amass this much information this quickly.

That said, it is important to emphasize that I did not use the Internet at large for verifying, and for the most part, absorbing any material. I stayed away from mass media, too, excepting the news sections of the professional journals and specialty publications, such as *Scientific American*. My intention was to translate, to the best of my ability, recent journal and professional literature into a single narrative for understanding the interrelatedness of the issues that face us as a society and individuals, an *Über*subject I decided to call Carbon Science. The articles were supplemented with many interviews and books by other science writers and writer-scientists.

Many sherpas guided me incrementally through these mountains of science. Over three years I interviewed many scientists about their work, numbering in the hundreds. I would like to single out professors around the country who read chapters or generously made time to answer my questions.

When I apologized in advance to David Helfand of Columbia University for asking some stupid questions, he pointed out with ebullient humor that there are no stupid questions, only ignorant ones. David is a force to be reckoned with, the chairman of the Astronomy Department, founder of Columbia's Frontiers of Science curriculum, and adviser to British Columbia's Quest University, which opened in 2007. Columbia has always been my spiritual home, so it was a privilege to meet and receive guidance from other scientists there, including Wallace Broecker, John Hunt, Darcey Kelly, James Leighton, Koji Nakanishi, and Joseph Patterson, and David Rind at NASA's Goddard Institute for Space Studies. When I told Columbia College dean Austin Quigley I quit my job to write a book about how the world seems to work, he replied, "Oh! That's a very Columbia College thing to do!"

Stuart Schreiber and Jon Clardy of Harvard University were generous with material they use in their undergraduate class, B47: Molecules of Life. Jeffrey Bada of the Scripps Institute of Oceanography helped me orient my thoughts early in the research. Harry Gray of Caltech read several drafts about the physical chemistry of the carbon atom. The pivotal interview that first put me on a productive track in 2004 is chronicled in the last pages of chapter 12.

Several people stand out for the sheer mass of material they read and their copious comments on it. Before I acknowledge them, I'd like to invoke the science writer's disclaimer: any errors or misjudgments critics and readers find in this book are, of course, my own.

Charles Alexander, who I worked with at *Time*, provided meticulous comments on the entire manuscript. David Baltaxe attacked the manuscript with curiosity and rigor. Jag Bhalla, the last gentleman-scientist, fed me a steady diet of books and articles, and offered critical commentary. Elizabeth Daerr at the Center for Inquiry in Washington gave me feedback on the first half—when there was only half. Farhad Ebrahimi reinforced the need for communicating what professionals don't call the "scientific method" but don't have any other name for either. Myer Ezrin and I spent many family gatherings huddled in corners, drawing circles and lines on napkins. Gabriel Eichler patiently answered many

RNA-related questions over many dinners. David Goldston weighed in with pivotal comments on several chapters. Dudley Herschbach provided me with feedback and reinforcement on a late draft. I was lucky to meet Dee Jacobs, a chemist and science writer, who walked me through the "dense forest jungle" of organic chemistry in chapters 8 and 9. Jennifer Cohen Oko and Elaine Heinzman helped kick an early draft into shape. Jennifer Ouellette provided critical feedback on a late draft. Allen Packman tore through the draft with knowledgeable eyes. Daniel Roston, a first cousin, is currently a Ph.D. candidate in chemistry at the University of Iowa. His perspicacious comments on every chapter separated much wheat from chaff. Christopher Tudge, who coordinates American University's undergraduate science curriculum and math program, read the manuscript and patiently walked through it with me.

Many other experts and friends looked over individual chapters for me—sometimes more than once—and provided instrumental feedback, or helped out in other ways: Mario Aguilera, Emily Kennedy Anthes, Frances Arnold, Bruce Becker, Ruth Greenspan Bell, Robert Berner, Michael Bernstein, Dennis Bramble, Jochen Brocks, Christopher Brodie, Jim Cleaves, James Collins, Felicia Day, Phil DeCola, Peter Del Tredici, Juliet Eilperin, James Fay, James Ferris, Woody Fischer, Tim Fitzgerald, Danny Franklin, Robert Freilich, Susanne Garvey, John Gowdy, Arthur Glasfeld, Andrew Goldstein, Susan Golden, Neil Gussman, Robert Hazen, Janet Hodur, Stuart Jordan, Jim Kasting, Sheril Kirshenbaum, Bill Klemperer, Kevin Knobloch, Andrew Knoll, Ray Kopp, James Leighton, Michael Lemonick, Karen Masterson, the Marco 8.5, Tina McDowell, Kevin O'Connor, Billy Pizer, Anthony Poole, Vicky Rainert, Mike Ranen, Amanda Ripley, Andy Ridgwell, David Robinson, Joseph Romm, Dana Royer, Bill Saporito, Daniela Schmidt, Roger Summons, Gwynedd A. Thomas, Penelope Trunk, Toby Tyrrell, Bruce van Voorst, Suzy Wagner, Olivia Walling, Stan Wellborn, Joseph Wolin, Brian Yourish, Jane Yourish, Stu Yourish.

No book is complete until it's been read and discussed. I can be reached at e@thecarbonage.com.

I wrote this book during my yearlong stay as an independent visiting scholar at Resources for the Future (RFF), in Washington, D.C. RFF is an influential, nonpartisan think tank, where some of the nation's top economists develop policies that address pressing energy and environmental issues. It was a research institution long before there were think tanks, and it retains the professionalism and analytical rigor that too often go missing in the nation's capital.

My years at *Time* magazine provided the backdrop against which I discovered slowly that nobody had written this book. In New York, it was a pleasure to work for Bill Saporito and Dan Goodgame, who served as business editors during my tenure there. As a member of the magazine's Washington Bureau, under the direction of Michael Duffy, James Carney, and Matthew Cooper, I had an opportunity to cover many pivotal events of recent years, a breadth of stories that encouraged me to tease out the connective tissue linking all of them. A short April 2001 article, in a business-section supplement, about the Santa Fe Institute introduced me to complexity science. Several great books have been written either about the institute or how complex adaptive systems emerge throughout nature. *The Carbon Age* is not about complexity per se. But living chemistry and evolution, whether on the level of a cell or the global economy, are the most powerful manifestations of emergent complexity.

The Swedish Embassy in Washington, D.C., provided a research trip to meet scientists and policy makers in Stockholm and visit the nation's Arctic region, in January 2007.

Many and varied thanks to Nicole St. Clair.

Lindy Amos drew my attention to the TAI Group's Power and Presence public communication seminars, where I learned to talk about what I'm talking about. Tony Arpante finds photographs in people like no one else.

Special honors to Shane Travis and Viveca Novak for plying me with C_2H_6O, and to Laura Freilich for the molecular Tinkertoys, which I broke open during chapter 8 to build a model of ginkgolide B.

My agent, David McCormick, gave me the encouragement and backstop I needed over the two years it took to research and structure a proposal. Walker & Company made the perfect home. George Gibson's passion for books and ideas knows no limits and he brings a contagious enthusiasm to his every project. It was a particular privilege to work with Jacqueline Johnson. Many thanks also to publicist Laura Keefe. Paula Cooper is my copyediting hero. Liz Peters made the production trains run smoothly and on time.

Without diminishing anyone else's role, I'd like to point out that my father, Jay Roston, read, I think, every page of every draft proposal, chapter, and manuscript generated over three years. Not to be outdone, my mother, Bonnie Roston, watched the media like a hawk for material relevant to *The Carbon Age*. My brother, Joel Roston, remains one of the most intriguing and startling people you might ever come across. His steady quips and insight kept me going through long days and nights of research. Jordyn Bonds is an equal and opposite creative force, speaking wit and wisdom far beyond her height.

Before I met my wife, Karen Yourish, life was a Precambrian moonscape. She showed up and suddenly there was enough oxygen to breathe and trilobites scampered about the continental shelf. Karen made this book possible in more ways than I can enumerate, from rigorously editing chapters, to patiently stepping over tall stacks of papers and books (often with me underneath them), to the continuous three-year-long flow of wisdom, support, and guidance that kept me going and keeps me going. I am the luckiest.

This book, being in part about how much faster human ideas evolve than humans themselves do, was produced in five times the span it took Karen to carry our daughter, Madeline Libby, who brought 7 pounds and, 15 ounces of insight about the origin of life into the last months of revisions. Quoth Buckminster Fuller:

> All universities have been progressively organized for ever finer specialization. Society assumes that specialization is natural, inevitable, and desirable. Yet in observing a little child, we find it is interested in everything and spontaneously apprehends, comprehends, and co-ordinates an ever expanding inventory of experiences. Children are enthusiastic planetarium audiences. Nothing seems to be more prominent about human life than its wanting to understand all and put everything together.

NOTES

Author's Note

Michael J. Benton and Richard J. Twitchett, "How to Kill," 358.

Prologue

1. National Research Council, *High Performance Structured Fibers*, 51.
2. Encyclopedia Britannica online, "Chemical Element: Abundances in Earth's Crust."
3. Atkins, *Atoms, Electrons and Change*, 19.
4. Campbell et al., *Biology*, 32.
5. Tanford and Reynolds, *Nature's Robots*, 3.
6. Ezrin, "Plastics Failure/People Failure," 17. Ezrin is a relative of the author.
7. Ogg, "Status of Divisions" 185.
8. Freese, *Coal*, 18; Lane, *Oxygen*, 76.
9. Klemme and Ulmishek, G. F. "Effective Petroleum Source Rock," IPCC, *Physical Science Basis*, 139.
10. Houghton "Balancing," 314.
11. Jackson, interview with the author, January 16, 2007.
12. Hazen, e-mail to the author, October 24, 2007.
13. Berner, "Long–Term Carbon Cycle," 323; IPCC, *Physical Science Basis*, 465.
14. IPCC, *Physical Science Basis*, 107.
15. Berner, *Phanerozoic Carbon Cycle*, 3–9.
16. Ibid., 6.
17. Erickson, "Gigantism," 772.
18. Niele, *Energy*, 98.

Chapter 1

1. Tyson and Goldsmith, *Origins*, 26.
2. Herbst, "Chemistry of Star-Forming Regions," 4017.
3. Atoms look nothing like the cover image of this book, a mini "solar system." Ernest Rutherford, an early twentieth-century British physicist, proposed this model of the atom in 1908, but colleagues never accepted it. The "Rutherford atom" shows negatively charged electrons whizzing about a positively charged nucleus, but makes no explanation for why they shouldn't "fall in" to the nucleus. In his 1955 essay, "The Myth of Sisyphus," 19–20, Albert Camus memorably takes on the Rutherford atom as a sign of the limits of knowledge:

   > Of whom and of what indeed can I say: "I know that!" [A]ll the knowledge on earth will give me nothing to assure me that this world is mine. You describe it to me and you teach me to classify it. You enumerate its laws and in my thirst for knowledge I admit that they are true. You take apart its mechanism and my hopes increase. At the final stage you teach me that this wondrous and multicolored universe can be reduced to the atom and that the atom itself can be reduced to the electron. All this is good and I wait for you to continue. But you tell me of an invisible planetary system in which electrons gravitate around a nucleus. You explain this world to me with an image. I realize then that you have been reduced to poetry: I shall never know. Have I the time to become indignant? You have already changed theories. So that science that was to teach me everything ends up in a hypothesis, that lucidity founders in metaphor, that uncertainty is resolved in a work of art. [Y]ou give me the choice between a description that is sure but that teaches me nothing and hypotheses that claim to teach me but that are not sure.

4. Tyson and Goldsmith, *Origins*, 27.
5. A study published as this book went to press posits the first stars may not have shone, per se, consisting mostly of dark matter. Spolyar, Freese, and Gondolo "Dark Matter."
6. Helfand, "History of the Stars and Elements," presented at Frontiers of Science lecture, New York, September 25, 2006.
7. Bethe, "Energy Production in Stars," 227.
8. The Sun's heat does not grow or fade substantially on human timescales and cannot explain modern global warming. See IPCC *Physical Science Basis*, 136 and 188–193. The most recent IPCC assessment revised the small impact of solar changes downward, to about 7.5 percent of the total warming.

9. Laurence, "Endless Duel of Atoms," 1.

10. Campbell et al., *Biology*, 35; Angier, *The Canon*, 97.

11. Bethe, "My Life," 1.

12. Bethe, "My Life," 3; Salpeter, interview with the author, May 25, 2006.

13. Wallerstein, "Synthesis of the Elements," 995–99.

14. Laurence, "Endless Duel of Atoms," 1.

15. Walling, "Research at the Kellogg," (Ph.D. dissertation, University of Minnesota, 2005), 112.

16. The Fahrenheit and Celsius temperature scales are pegged to the behavior of water on Earth. They tell us whether to pack a sweater or bathing suit, but are of limited use in describing stellar heat or interstellar cold. The kelvin scale is used for measuring temperatures outside the range of human experience. There is no negative part of the scale. Zero degrees kelvin is absolute zero, the absence of heat. One degree kelvin and one degree Celsius have the same magnitude; zero degrees kelvin is –273 degrees Celsius and zero degrees Celsius is 273 degrees kelvin.

17. J. Richard Gott III has written: "Predicting that the radiation existed and then getting its temperature correct to within a factor of 2 was a remarkable accomplishment—rather like predicting that a flying saucer 50 feet in width would land on the White House lawn and then watching one 27 feet in width actually show up. One could call it the most remarkable scientific prediction ever to be verified empirically." Gott, *Time Travels*, 160; quoted in Tyson and Goldsmith, *Origins*, 57.

18. Salpeter, interview with the author, May 25, 2006; Gregory, *Fred Hoyle's Universe*, 47.

19. Walling, "Research at the Kellogg," (Ph.D. dissertation, University of Minnesota, 2005), 114.

20. Wallerstein, "Synthesis of the Elements," 1020.

21. Walling, "Research at the Kellogg," (Ph.D. dissertation, University of Minnesota, 2005), 110–11.

22. Hoyle, *Home Is Where the Wind Blows*, 264.

23. Barrow, *Constants of Nature*, 153–54.

24. Hoyle, *Home is Where the Wind Blows*, 265.

25. Fynbo, et al., "Revised Rates," 136–37; Fynbo, interview with the author, May 8, 2007.

26. Fred Hoyle is one of the twentieth century's greatest gifts to historians of astrophysics. Everyone I encountered who encountered him had a story to tell, usually about Hoyle's later years, when he became a proponent of ideas so far removed from physical evidence and logic that he lost credibility. Jeffrey Bada recalled how he and Stanley Miller, of the Scripps Institution of

Oceanography, couldn't contain their laughter during a Hoyle lecture on possible origins of life. Lewis Snyder, an astronomer at the University of Illinois, observed drily that Hoyle enjoyed a good show. Harry Kroto exclaimed, "He called me a liar on television!" Edwin Salpeter is extremely self-effacing about his pivotal role in the discovery of the triple-alpha process—alternatively known as the Salpeter process. I asked him if Hoyle's prediction was so significant, why wasn't the process named after him instead? "That's just the nickname other people give to it," Salpeter said. "Hoyle figured it out. Let me put it this way, I'm a more pleasant guy than Fred Hoyle. Maybe they liked me better. He was a slightly difficult guy to get along with. But a real genius." Two Hoyle biographies have come out in the last few years, Simon Mitton's *Conflict in the Cosmos* and Jane Gregory's *Fred Hoyle's Universe*. Hoyle wrote an autobiography, and a short autobiographical essay, both listed in the bibliography. His 1959 novel, *The Black Cloud*, holds up as an intriguing and fun sci-fi read.

27. Hoyle "Universe," 15–16.
28. Levi, *The Periodic Table*, 230.
29. Feynman, *Six Easy Pieces*, 4.
30. Ehrenfreund and Charnley, "Organic Molecules," 451–5; Kroto, interview with the author, July 6, 2006; Henning and Salama. "Carbon in the Universe," 2204–5.
31. Tyson and Goldsmith, *Origins*, 113; Seeds, *Foundations*, 258.
32. Herbst, "Chemistry of Star-Forming Regions," 4018–21.
33. Herbst, "Chemistry of Interstellar Space," 171.
34. Ouellette, e-mail to the author, November 17, 2007.
35. Seeds, *Foundations*, 204.
36. Roth et al., "Interstellar Cyanogen," L67.
37. Klemperer, "Chemistry of Interstellar Space."
38. Herschbach, e-mail to the author, November 13, 2007.
39. Wilson, Jefferts, and Penzias, "Carbon Monoxide in the Orion Nebula," L43; Klemperer, "Chemistry of Interstellar Space."
40. Travers et al., "Laboratory Detection," L61.
41. Nelson and Cox, *Lehninger Principles*, 116.
42. Snyder et al., "A Rigorous Attempt," 914.

Chapter 2

1. National Research Council, "The Limits of Organic Life," 6.
2. National Research Council, "Assessment of the NASA Astrobiology Institute" 5.

3. Szostak et al., "Synthesizing Life," 390. This is a useful phrase, but a loaded one. A "black box" in Szostak's usage signifies a great, probably the greatest scientific puzzle. That this phrase was adopted by a proponent of the intelligent design restatement of creationism is by now a historical footnote.

4. Waldrop, "How Do You Read," 578.

5. Waldrop, "Did Life Really Start," 1248.

6. Szathmáry "In Search Of," 469–70.

7. Belief that the Earth is only several thousand years old carries a curious implication. The physical evidence for the Earth's age emerged from the same atomic discoveries that later gave the world nuclear weaponry and power plants. The scientific understanding of uranium isotopes that produce the date 4.5.billion years ago is the same understanding of uranium isotopes that led to the production and detonation of nuclear bombs. If scientists do not understand uranium decay well enough to date the Earth, there also cannot be, and can never have been, nuclear weaponry. Certainly a world and a history absent these weapons are desirable, but they are counterfactual.

8. Bada, "Origins of Life," 98.

9. Valley, "A Cool Early Earth?" 60.

10. Encyclopedia Britannica online, "Chemical Element: Abundances in Earth's Crust."

11. Chemical Abstracts Service, Registry Number and Substance Counts. http://www.cas.org/cgi–bin/cas/regreport.pl, (accessed December 16, 2007); Clardy, e-mail to the author, March 12, 2007. Several definitions of "organic chemistry" occur: anything with carbon in it, anything with carbon-hydrogen bonds in it, the professional discipline of synthesizing carbon compounds. For our purposes, I choose the first definition.

12. Morowitz and Smith, "Energy Flow," 1.

13. Chyba et al., "Cometary Delivery," 366.

14. IPCC, *Physical Science Basis*, 749.

15. Grinspoon, *Lonely Planets*, 91–93.

16. Valley "A Cool Early Earth?" 60; Simon, "Evidence From Detrital Zircons," 177.

17. Darwin, Francis, ed., *Life and Letters of Charles Darwin*, 18.

18. Sephton, "Organic Compounds," 292.

19. Gold, "Deep, Hot Biosphere," 6045.

20. Orgel, interview with the author, May 10, 2006. Orgel passed away in late 2007.

21. Ausubel, e-mail to the author, November 25, 2007.

22. Wills and Bada, *Spark of Life*, 35–40.

23. Bada, "Origins of Life," 100.

24. Bada, interview with the author, LaJolla, California, January 31, 2006.

25. Kump et al., *Earth System*, 235.
26. Chyba, "Rethinking Earth's Early Atmosphere," 962–63; Trainer et al., "Organic Haze," 18036–38.
27. Deckert et al., "Complete Genome," 353.
28. Pollack, "Emergence of Information in the Universe," (Frontiers of Science lecture, October 30, 2006).
29. Charlebois and Doolittle, "Computing," 2469.
30. Banton et al., *Evolution*, 53.
31. Gilbert, "Origin of Life," 618.
32. Cech "Enhanced," 878.
33. Hoelzer, Smith, and Pepper, "On the Logical Relationship," 1793.
34. Rode, et al., "Strong Paramagnetism," 298.
35. National Research Council, *Limits of Organic Life*, 19.
36. David Deamer, preface to Hazen, *Genesis*, ix.
37. Deamer, "Origins of Membrane Structure," 67–79.
38. Hazen, "Mineral Surfaces," 1715.
39. Hazen, *Genesis*, 157.
40. Ferris, "Mineral Catalysis," 145.
41. Ferris, "Mineral Catalysis," 145; Ferris "Montmorillonite-catalysed," 1780.
42. Ferris, "Mineral Catalysis," 149.
43. Szostak, interview with the author, June 12, 2007.
44. Barton et al., *Evolution* 10–11.

Chapter 3

1. De Duve "Birth of Complex Cells," 56.
2. Knoll, *Life on a Young Planet*, 107.
3. Not until 2007 did astronomers even confidently detect molecular oxygen in space, inhabiting a dense cloud in the Ophiuchus constellation, about five hundred light years away; Larsson et al., "Molecular Oxygen," 1–11.
4. Kasting, e-mail to the author, September 4, 2007.
5. Bernt, Climate Action 2007 conference, Carnegie Institution for Science, Washington, D.C., October 22, 2007.
6. Ditty, Williams, and Golden, "Cyanobacterial Circadian Timing Mechanism," 515.
7. Catling and Claire, "Loss of," 3; Azam and Worden, "Microbes, Molecules," 1622.
8. Jones, "Personal Effects," 14–16.
9. Gaylarde, Silva, and Warscheid, "Microbial Impact," 342.

10. Sommers, et al., "Freshwater Ferromanganese," 407.
11. Schopf, "Fossil Evidence," 880; Brasier et al., "Fresh Look," 887; Allwood et al., "Stromatolite Reef," 714.
12. NASA Ames Research Center, "Stromatolite Explorer," multimedia Web site, http://microbes.arc.nasa.gov/movie/large–qt.html (accessed November 26, 2006).
13. Whitman, Coleman, and Wiebe, "Prokaryotes," 6578.
14. Azam and Worden, "Microbes, Molecules," 1622–23.
15. Todar, "Bacterial Flora of Humans."
16. Oren, "A Proposal for Further Integration," 1895.
17. Whitman, Coleman, and Wiebe "Prokaryotes," 6582.
18. Grinspoon. *Lonely Planets*, 119.
19. Brocks, interview and e-mail to the author, July 23, 2006, February 7, 2007, April 25, 2007.
20. Golden and Canales, "Cyanobacterial Circadian Clocks," 191–93; Golden, interview with the author, November 28, 2006.
21. Nelson and Cox, *Lehninger Principles*, 694.
22. Krauss, *Atom*, 206.
23. Nelson and Cox, *Lehninger Principles*, 748.
24. Campbell et al., *Biology*, 200.
25. Shallenberger, *Taste Chemistry*, 153.
26. Beerling, *Emerald Planet*, 175.
27. Calvin, *Following the Trail of Light*, 134.
28. Hofmann, "Precambrian Microflora," 1040.
29. Darwin, *Origin of Species*, 269.
30. Ruddiman *Plows, Plagues and Petroleum*, 15.
31. Schopf, "Fossil evidence," 869–70.
32. eBay, "Huge 3.5 inch Stromatolite," item number 130051673910, December 3, 2006.
33. Brocks and Pearson, "Building the Biomarker," 248; Summons et al., "2-Methylhopanoids," 554.
34. Knoll, "New Molecular Window," 1025–26.
35. Brocks, e-mail to the author, February 7, 2007.
36. Summons, e-mail to the author, November 9, 2007.
37. Summons, e-mail to the author, October 15, 2007.
38. Eigenbrode and Freeman, "Late Archean Rise," 15759.
39. Goldblatt, Lenton, and Watson, "Bistability of Atmospheric Oxygen," 683.
40. Kopp et al., "Paleoproterozoic Snowball Earth," 11133.
41. Kasting "Rise of Atmospheric Oxygen," 819; Holland "Oxygenation of the Atmosphere and Oceans," 905.

42. Catling and Claire, "How Earth's Atmosphere Evolved," 11.

43. Kopp et al., "Paleoproterozoic Snowball Earth," 11131.

44. Ridgwell and Zeebe, "Global Carbonate Cycle," 302.

45. Levi, *The Periodic Table*, 226.

46. Hay, "Tectonics and Climate," 409.

47. Des Marais, "Carbon Isotope Evidence," 607.

48. Kirschvink, "Red Earth," 19.

49. Lane, *Oxygen*, 48; Kirschvink, "Red Earth," 16.

50. Kopp et al., "The Paleoproterozoic Snowball Earth," 11132; Kirshvink "Red Earth," 15.

Chapter 4

1. Holland, "Oxygenation of the Atmosphere and Oceans," 908; Buick, Des Marais, and Knoll, "Stable Isotopic Composition," 153; Ogg "Status of Divisions," 194.

2. Tyrrell and Young, "Coccolithophores," (forthcoming).

3. The notion of "inorganic carbon" might seem oxymoronic since "inorganic chemistry" is defined as being carbon free. In fact, within geochemistry, "inorganic carbon" describes material that rides through the Earth system on physical forces, unmediated by biology or at least without passing through soft issue.

4. Fortey, *Trilobite*, 92.

5. Dawkins and Krebs, "Arms Races," 489.

6. Knoll et al., "A New Period," 621.

7. Ogg, "Status of Divisions," 185.

8. Amthor et al., "Extinction of Cloudina," 431.

9. Zeebe and Westbroek "Simple Model," 1.

10. Hsu et al., "Strangelove Ocean," 809.

11. Ridgwell and Zeebe, "Global Carbonate Cycle," 1309; Hayes, interview with the author, July 23, 2006.

12. MacMennamin and MacMennamin, *Emergence of Animals*, 97–103.

13. Ward, *Out of Thin Air*, 12.

14. Kerr, "Shot of Oxygen," 1529; Canfield et al., "Late-Neoproterozoic Deep-Ocean," 92.

15. Margulis. *Environmental Evolution*, 141.

16. Nelson and Cox, *Lehninger Principles*, 38–39; Lane, *Power, Sex, Suicide*, 56–61.

17. Falkowski et al., "Modern Eukaryotic Phytoplankton," 358.

18. Knoll, "Biomineralization and Evolutionary History," 331.

19. Marin et al., "Skeletal matrices," 1554.

20. Knoll, "Biomineralization and Evolutionary History," 332, 339.

21. Ibid., 330.

22. Bengtson and Zhao, "Predatory Borings," 367.

23. Hua, Pratt, and Zhang, "Borings in *Cloudina* Shells," 457.

24. Knoll and Carroll, "Early Animal Evolution," 2129; Knoll, *Life on a Young Planet* 8–11, 14.

25. Bengtson, "Origins and Early Evolution of Predation," 300.

26. John Hayes, interview with the author, July 23, 2006.

27. Nelson and Cox, *Lehninger Principles*, 1115.

28. Bengtson, "Origins and Early Evolution of Predation," 300.

29. Huntley and Kowalewski, "Strong Coupling," 15006.

30. Benton and Twitchett, "How to Kill," 358–62.

31. Knoll, "Biomineralization," 342–44.

32. Falkowski et al., "Modern Eukaryotic Phytoplankton," 359.

33. Tyrrell and Young, "Coccolithaphores," (forthcoming), McPhee "Season on the Chalk,"; Huxley, "On a Piece of Chalk," 174–76.

34. Rost and Riebesell, "Coccolithaphores and the Biological Pump," 106.

35. Young and Henriksen, "Biomineralization Within Vesicles," 200–205; Tyrrell, e-mail to the author, March 19, 2007.

36. Tyrell, e-mail to the author, January 31, 2008.

37. Shatto and Slowey, "Thinking Big."

38. Kolbert "Darkening Sea," 68.

Chapter 5

1. John Lough, interview with the author, Chicago, Illinois, April 2006.

2. Ishikawa and Swain, *Hiroshima and Nagasaki*, 87; Del Tredici, "Ginkgo and People," 4.

3. Darwin, *Origin of Species*, 105.

4. Desmond and Moore, *Darwin,* 456–61.

5. Delsuc et al., "Phylogenomics," 361–75.

6. Simonson et al., "Decoding the Genomic Tree of Life," 6611; Brown, "Ancient Horizontal Gene Transfer," 131.

7. Deep Green, data matrices available at http://ucjeps.berkeley.edu/bryolab/GPphylo/data_matrices.php, (accessed February 2007).

8. Green Plant Phylogeny Research Center, "Hyperbolic Trees," http://ucjeps.berkeley.edu/map2.html (accessed February 20, 2007).

9. Greb et al., "Evolution and Importance," 4.

10. Beerling et al., "Evolution of Leaf-Form," 352–54.

11. Greb et al., "Evolution and Importance," 4.

12. Chameides and Perdue, *Biogeochemical Cycles*, 64.

13. Knoll, "Geological Consequences," 10.

14. Berner, "The Carbon Cycle and CO_2," 77–78; Royer, e-mail to the author, October 19, 2007.

15. Berner, *Phanerozoic Carbon Cycle*, 1–10.

16. Greb et al., "Desmoinisian Coal Beds," 130–31.

17. Cole et al., "Organic Geochemistry," 1441–42.

18. Klemme and Ulmishek, "Effective Petroleum Source Rocks."

19. Afifi, "Ghawar"; Afifi, e-mail to the author, June 6, 2007.

20. Stoneley, "Review of Petroleum," 266.

21. Beerling "Evolution of Leaf-Form," 354.

22. ITIS Standard report page, "Ginkgo biloba."

23. Del Tredici, "Evolution, Ecology and Cultivation," 9.

24. Tralau, "Evolutionary Trends," 64.

25. Raver, "Hardy Ginkgo Trees," 34.

26. Tralau, "Evolutionary Trends," 64.

27. Ibid., 68–70.

28. Zhou and Zheng, "Missing Link," 821.

29. Del Tredici. "Evolution, Ecology and Cultivation," 7–9.

30. Ibid., 13.

31. Del Tredici "Ginkgo and People," 3.

32. Royer et al., "Ecological Conservatism," 90.

33. Del Tredici "Ginkgo in America," 158

34. Friedman, interview with the author, September 13, 2006.

35. Norstog, Gifford, and Stevenson, "Comparative Development," 9.

36. Retallack, "Carbon Dioxide and Climate," 660.

37. Beerling and Royer "Reading a CO_2.signal," 86, 90–94.

38. Chen et al., "Assessing the Potential," 1309–13.

39. Beerling and Royer "Fossil Plants as Indicators," 544.

40. Retallack, "Carbon Dioxide and Climate," 665.

41. IPCC, *Physical Science Basis*, 440.

42. Royer et al. "Climate Sensitivity Constrained," 2310.

43. Matsumoto et al., "Climate Change and Extension," 1634.

44. Del Tredici "Phenology of Sexual Reproduction," 268.

45. Hendricks, e-mail to the author, November 22, 2005.

Chapter 6

1. Kittler et al., "Molecular Evolution of *Pediculus humanus*," 1414.
2. Bramble and Lieberman, "Endurance Running," 345–52.
3. Galik et al., "External and Internal Morphology," 1450.
4. Berge and Daynes, "Modeling Three-Dimensional Sculptures," 149–50.
5. Bramble and Lieberman, "Endurance Running," 346.
6. Howell, *Early Man*, 41-45.
7. Gould, *Wonderful Life*, 27–35.
8. Bramble, e-mail to the author, November 29, 2007; Chen "Born to Run," 65.
9. Bramble and Lieberman, "Endurance Running," 345.
10. Nelson and Cox, *Lehninger Principles*, 490.
11. Landa, "Oink If You Love Coal," 60.
12. Lane, *Power, Sex, Suicide*, 26.
13. Ibid., 25–26.
14. Forte, "Bio 1a: General Biology at UC–Berkeley."
15. Krebs, "The Citric Acid Cycle," 407.
16. Lane, *Power, Sex, Suicide*, 85.
17. Mitchell, "David Keilin's Respiratory Chain Concept," 295.
18. Heinrich, *Racing the Antelope*, 122.
19. Carrier, "Energetic Paradox," 483.
20. Ibid., 485; Bramble, e-mail to author, November 29, 2007.
21. Taylor and Rowntree, "Temperature Regulation," 850.
22. Cairns–Smith, "Sketches," 157.
23. Venables, et al., "Determinants of Fat Oxidation," 164; Helge, "Muscle Fat Utilization," 1255.
24. Nelson and Cox *Lehninger Principles*, 363.
25. Ibid., 368.
26. Carrier, "Energetic Paradox," 487.
27. National Research Council, *Limits of Organic Life*, 20.
28. Greene, "That Famous Equation," 31.
29. Smil, *Energy*.
30. Ibid., 8–9.
31. Lim, interview with the author, January 4, 2008.
32. Nelson and Cox *Lehninger Principles*, 844.
33. Wilson and Nicoll, "Endocannabinoid Signaling," 678.
34. Le Couteur and Burreson, *Napoleon's Buttons*, 265.
35. Knight, "No Dope," 114–15; Aguilera et al., "Detection of Epitestosterone Doping," 629–36.

36. LaBarbera, "Why Wheels Won't Go," 395–408.

37. Barton et al., *Evolution*, 319.

38. Wilson, "Bicycle Technology," 82.

39. Ibid., 88.

Chapter 7

1. Hoffmann, *Same and Not the Same.* The theme of the natural and unnatural, of nature and man, is one of the oldest in all history in literature. My decision to split *The Carbon Age* in half, along these lines, came upon considering two essays in Hoffmann's book, pages 107–15.

2. Mokyr, "Useful Knowledge," 2.

3. Even Dawkins has some fun at their expense in *The Selfish Gene.* "I have been a bit negative about memes, but they have their cheerful side as well."

4. Arrow, interview with the author, March 30, 2007. For a more mainstream economist's perspective on evolution and economics, see also Krugman, "What Economists Can Learn."

5. Endy, "Foundations of Synthetic Biology," 451.

6. Benyus, *Biomimicry*, 97.

7. Pollack, "Emergence of Information."

8. Diamond, *Third Chimpanzee*, 143.

9. Fujimoto, *Competing*, 5, 40.

10. U.S. Department of Energy, *Basic Research Needs*, 29.

11. Hogg-Misra et al., "Revised Hazy Methane Greenhouse."

12. Pyne, *Fire*, xvi.

13. Klein and Edgar, *Dawn of Culture*, 155.

14. Klein, *Dawn of Culture*, 157.

15. Niele, *Energy*, 45–46.

16. Klein, *Dawn of Culture*, 144.

17. Kirsch, *Electric Car*, 11.

18. Hillier and Coontes, "Hillier's Fundamentals," 32.

19. Ibid., 31–32.

20. Yergin, *Prize*, 80, 209.

21. Kirby et al., *Engineering in History*, 405–407.

22. Womack et al., *Machine*, 23.

23. Ibid., 24.

24. Ibid., 30.

25. Ibid., 29.

26. U.S. Department of Energy, *Basic Research Needs*, viii.

27. Fujimoto, *Competing*, 66.
28. Ibid., 57–59.
29. Ibid., 70.
30. Tierney, "Ford Slams Toyota."
31. Atkins, *Atoms, Electrons and Change*, 61–62.
32. Ibid., 59, 106.
33. U.S. Department of Energy, *Basic Research Needs*, 37–38.
34. Bloomfield, "Working Knowledge," 108.
35. Dukes "Burning Buried Sunshine," 31–44.
36. Reyes and Sepúlveda "PM–10 Emissions," 1714–19; Smil, "Energy," 14.
37. International Energy Agency, *Energy Technologies*, 39; Smil, "Energy," 14.
38. Aldrich et al., "In Defense of Generalized Darwinism," (forthcoming).
39. Smith and Szathmáry, *The Origins of Life*, 107; cited in Lane, *Power, Sex, Suicide*, 111.

Chapter 8

1. Wilson and Danishefsky, "Small Molecule Natural Products," 8329–31.
2. Michel-Zaitsu, "On Engelbert Kaempfer's Ginkgo."
3. Nakanishi, *A Wandering Natural Products Chemist*, 60.
4. Nakanishi, "Terpene Trilactones from Ginkgo biloba," 4984.
5. Ibid.
6. Koltai et al., "Effect of BN 52021," 135–36.
7. Nakanishi, *A Wandering Natural Products Chemist*, 58–68.
8. Corey "Logic of Chemical Synthesis," 686; Baran, interview with the author, July 12, 2006; Woodward, "Arthur C. Cope Award Lecture," 4270.
9. Hazen, *Genesis*, 134.
10. Borek, *Atoms Within Us*, 4–5.
11. Ramberg and Somsen, "Young J. H. van 't Hoff," 51.
12. Nicolaou et al., "Art and Science," 1241–42.
13. Ibid., 1,233–34; McMurry, *Organic Chemistry*, 583.
14. Joseph Patterson is a professor of astronomy at Columbia University. He expressed this divide clearly. "I was never interested in chemistry. Even in high school I never took it. I always thought chemistry was just this weird thing that happens because of the Pauli Exclusion Principle."
15. Brock, *Chemical Tree*, 569.
16. Barton, "Principles of Conformational Analysis," 300.
17. Bartlett et al., "Robert Burns Woodward," 585–87; *Time*, July 18, 1960.
18. Todd and Conforth, "Robert Burns Woodward," 77.

19. Corey and Cheng, *Logic of Chemical Synthesis*, 2.
20. Del Tredici, "Evolution, Ecology, and Cultivation," 20–21.
21. Nakanishi, "Terpene Trilactones from Ginkgo biloba," 4988–89; Wang and Chen, "Ginkgolide B, a constituent of Ginkgo biloba," 141, 146–148; Nakanishi , interview with the author, New York, January 16, 2007.
22. Montgomery, "*Ginkgo biloba* Dietary Supplement."

Chapter 9

1. Kelly, *Gunpowder*, 8, 15.
2. Thomas, e-mail to the author, May 30, 2007.
3. Principe, "Chemistry," 153.
4. Kelly, *Gunpowder*, ix, 23.
5. Ibid., 117.
6. Ibid., 118.
7. Buchanan, *Gunpowder, Explosives and the State,* 4.
8. Kelly, *Gunpowder*, 6.
9. Ibid., 6, 231.
10. Rice, Smokeless Powder," 355.
11. Kirschvink and Hagadorn, "Grand Unified Theory," 139.
12. Kelly, *Gunpowder;* Christine O'Brien, e-mail to the author, January 2, 2008.
13. Harris and Paxton, *Higher Form of Killing*, 11.
14. Fenichell, *Plastics*, 155; Mokyr, *Gifts of Athena*, 109; O'Brien, e-mail to the author, January 2, 2008.
15. Kwolek, "Innovative Lives"; Nelson and Cox *Lehninger Principles;* Ezrin, *Plastics Failure Guide*, 1–4.
16. Fenichell, *Plastics*, 154–61.
17. Fenichell, *Plastics*, 165–71; Kwolek, interview by Bernadette Bensaude-Vincent.
18. O'Brien, e-mail to the author, January 2, 2008.
19. Le Couteur and Burreson, *Napoleon's Buttons*, 91.
20. Kwolek, "Innovative Lives."
21. Ibid.
22. Tanner et al., "Kevlar Story," 650.
23. Ibid.
24. *Economist* 1985, "Profit-Proof Kevlar?" 88; O'Brien, e-mail to the author, January 2, 2008.
25. Hazen, *Diamond Makers*, 5–6.
26. Tennant, "On the Nature of the Diamond," 123.

27. Wagner, interview with the author, December 6, 2006.
28. Wotiz, *Kekulé Riddle*.
29. Kamminga, "Kekulé Riddle," 1463.
30. Magat, "Fibres," 61.
31. Shubin, interview with the author, July 31, 2006.
32. U.S. Department of Commerce, "Soft Body Armor," 1–2.

Chapter 10

1. Tweney, Faraday, and Gooding, "Michael Faraday's 'Chemical Notes,'" viii.
2. Brock, *Chemical Tree*, 150.
3. Ibid., 371.
4. Eliot, *Harvard Classics*, 174–75.
5. Freese, *Coal*, 37.
6. Hansen et al., "Dangerous."
7. Christianson, *Greenhouse*, 11–12.
8. Fleming, Historical Perspectives on Climate Change, 68.
9. Fleming, *Historical Perspectives on Climate Change*, 68–9; IPCC, *Physical Science Basis*, 103–105; Kolbert, *Field Notes*, 35–36.
10. Gray, *Braving the Elements*, 332.
11. Keeling, "Brief History."
12. Woodman, "Exact Estimation," 259–60.
13. Christianson, *Greenhouse*, 141; IPCC *Physical Science Basis*, 105.
14. IPCC, *Physical Science Basis*, 749.
15. Lovelock, "Travels with an Electron Capture Detector." Lovelock is best known for his part metaphor, part hypothesis that the Earth system can be thought of as a self-regulating organism. He chose the name "Gaia" to describe the integrated system, after the Greek goddess of the Earth.
16. Keeling, "Rewards," 33–34.
17. Ibid.
18. Christianson, *Greenhouse*, 155–56. Revelle's name received wide attention in 2006, fifteen years after his death, when former vice president Al Gore narrated his formative college experience studying with Revelle in the film, *An Inconvenient Truth*.
19. Smil, *Cycles of Life*, 100.
20. Houghton, "Balancing," 318.
21. Keeling, "Concentration and Isotopic Abundances," 201.
22. Weart, e-mail to the author, October 19, 2007; Keeling, "Rewards."
23. Keeling, "Concentration and Isotopic Abundances," 203.

24. Keeling "Rewards," 53.

25. Houghton, "Balancing," 313. As early as February 1965, the White House spoke publicly about the potential dangers of atmospheric change. Lyndon Baines Johnson wrote in a Special Message to the Congress, "Air pollution is no longer confined to isolated places. This generation has altered the composition of the atmosphere on a global scale through radioactive materials and a steady increase in carbon dioxide from the burning of fossil fuels."

26. IPCC, *Physical Science Basis*, 96.

27. Ibid., 96–97.

28. Kump et al., *Earth System*, 47.

29. Ibid.

30. Ibid.

31. IPCC, *Physical Science Basis*, 100.

32. Ausubel, e-mail to the author, November 25, 2007.

33. Houghton, "Balancing," 331.

34. Quoted in Romm, *Hell and High Water*, 11. An "ornery beast," or at least a stuffed toy serpent several yards long, winds around the ceiling of Broecker's office.

35. Hansen et al., "Climate Change and Trace Gases," 1928.

36. IPCC, *Physical Science Basis*, 749.

37. IPCC, *Impacts, Adaptation and Vulnerability*, 824.

38. Hansen, "Declaration of James E. Hansen."

39. Wing, interview with the author, Washington, D.C., September 15, 2006.

40. Stonely, "Review of Petroleum," 264.

41. Sebald, *The Rings of Saturn*, 170.

42. Tol, "Marginal Damage," 2065.

43. Weitzman, "Stern Review," 19.

44. Greenspan Bell, e-mail to the author, October 31, 2007.

45. *Stern Review*, Executive Summary, 10.

46. IPCC, *Impacts, Adaptation and Vulnerability*, 822. The difference between the price of a ton of carbon and a ton of carbon dioxide can be very confusing. Policymakers and carbon-market participants tend to use the metric ton of carbon dioxide (t/CO_2) as their standard unit of measurement and trade, because their primary concern is carbon in the main form it takes as an industrial emission, CO_2. But most carbon–cycle scientists prefer to measure greenhouse gas emissions in terms of tons of carbon (t/C). If they only measure the movement of tons of carbon around the world, it doesn't matter if it's in biomass as carbohydrates, the sky as CO_2, or in methane, oil or coal. Measuring in t/C makes it easier to think about carbon's molecular transformations and the movement of carbon through its cycle.

47. Tol, "Marginal Damage," 2073.
48. Gowdy, "Behavioral Economics," (forthcoming).
49. Pacala and Socolow, "Stabilization Wedges," 968; Field, "Alarming Acceleration."
50. Reisch, "From Coal Tar."
51. Vernadsky, "Few Words about the Noösphere."
52. Haberl et al., "Quantifying and Mapping," 12942.
53. Schmidt-Bleek, *Fossil Makers*, 20.
54. Palumbi, "Humans," 1786–90.
55. McRae, interview with the author, Cambridge, MA, September 29, 2006.
56. Rind, e-mail to the author, January 1, 2008.

Chapter 11

1. Gardner, "Shotgun Mapping of Transcription Regulation: The Hunt for Genetic Gadgetry," (presentation at the Synthetic Biology 2.0 conference, Berkeley, California, May 20–22, 2006). Accessed online, January 2008.
2. "Human Journey, Human Origins," *National Geographic Magazine*, March 2006, 60–69.
3. Ke, "African Origin," 1151.
4. Barton et al., *Evolution*, 452.
5. Behar et al., "Genographic Project," 1083.
6. Ibid., 1089.
7. Wells, "Out of Africa," 114.
8. Moore, "Cramming More Components."
9. Service, "$1,000 Genome," 1544.
10. Qian et al., "High Resolution," 259.
11. Several months later, Watson ignominiously suggested of people with (recent) African ancestry, "All our social policies are based on the fact that their intelligence is the same as ours—whereas all the testing says not really." A book on evolution put out some months before by Cold Spring Harbor Laboratories—where he resigned a leadership post after retracting the statement—notes that no one has discovered any correlation between "cognition and other sensitive traits" and geographical origin.
12. Collins, "Who's Afraid of Synthetic Biology?"
13. Bügl et al., "DNA Synthesis," 628.
14. Baker et al., "Engineering Life," 46.
15. Collins (Boston University), "Who's Afraid of Synthetic Biology?" (unpublished essay).

16. Levskaya, "Engineering," 441–42.
17. Rösler, "Hello World Collection."
18. Registry of Standard Biological Parts, "Part: BBa_I15010."
19. Compton and Bunker, "Genesis of Curriculum," 12.
20. Julia Moskin, "Creamy, Healthier Ice Cream? What's the Catch?" *New York Times*, July 26, 2006.
21. Chan et al. "Refactoring T7," E1–E2
22. Knight, "Engineering Novel Life," 1.
23. Ball, "What Is Life?"
24. Bügl et al., "DNA Synthesis," 627–29.
25. Bill Joy and Ray Kurzweill, "Recipe for Destruction." *New York Times*, October 17, 2005.
26. Gingeras, "Origin of Phenotypes," 682.
27. Gerstein et al., "What is a Gene?" 671.
28. Arnold, "Unnatural Selection," 43–44.
29. Arnold and Otey, "Libraries."
30. Nelson and Cox, *Lehninger Principles*, 783.

Chapter 12

1. Li–Hung et al., "Long-Term Sustainability," 479.
2. Tyson and Goldsmith, *Origins*, 244.
3. Herz, *Shape of Carbon Compounds*, 4.
4. Marchetti, "On Decarbonization."
5. Ausubel, "Energy and Environment"; Marchetti, "Nuclear Plants and Nuclear Niches"; Ausubel, "Decarbonization."
6. Energy Information Administration, "International Energy Outlook."
7. Schmitt, "Mining the Moon."
8. In his 2005 science fiction novel *Accelerando*, Charles Stross ends a passage on the imagined hyperprogression of technology decades into the twenty-first century with the punch line, "Fusion power is still, of course, fifty years away."
9. R. D. Johnson, and C. Holbrow, ed., *Space Settlements: A Design Study* http://www.nss.org/settlement/nasa/75SummerStudy/Chapt.1.html#History.
10. Vernadsky, "Few Words About the Noösphere."
11. Dyson, "Search for Artificial Stellar Sources," 1667.
12. Rutan, "Burt Rutan."
13. Haskin, "Water and Cheese."
14. Ibid.

15. Smil, *Energy at the Crossroads*, 350.
16. Anker, *Ecological Colonization*.
17. Ibid.; Fuller, *Operating Manual for Spaceship Earth*.
18. Anker, "Ecological Colonization," 250.
19. Gewin, "Discovery in the Dirt."
20. Smil, interview with the author, May 2007.
21. Gott, "Implications of the Copernican Principle," 319.
22. Avise, "Real Message."
23. Broad, "Too Rich a Soil."
24. Comparisons probably stop there, and rather abruptly.
25. Kroto, "C_{60}: Buckminsterfullerene," 115.
26. Curl et al., "How the News," 185.
27. Smalley, interview with the author, Houston, Texas, May 6, 2004.
28. Hoffmann, "How Should Chemists Think?," 72.
29. Zhang, Eaton, and Gilardi, "Hepta- and Octanitrocubanes," 402.
30. Stanely, "Nobel Just the First."

BIBLIOGRAPHY

Afifi, Abdulkader. "Ghawar: The Anatomy of the World's Largest Oil Field." AAPG Distinguished Lecture, Search and Discovery Article #20026, 2005. Accessed from http://www.searchanddiscovery.net/documents/2004/afifi01/index.htm on June 11, 2007.

Aguilera, Rodrigo, Caroline K. Hatton, and Don H. Catlin. "Detection of Epitestosterone Doping by Isotope Ratio Mass Spectrometry." *Clinical Chemistry* 48:4 (2002): 629–36.

Aldersey-Williams, Hugh. *The Most Beautiful Molecule: The Discovery of the Buckyball.* New York: John Wiley & Sons, 1995.

Aldrich, Howard E., Geoffrey M. Hodgson, David L. Hull, Thorbjørn Knudsen, Joel Mokyr, and Viktor J. Vanberg. "In Defense of Generalized Darwinism." (Forthcoming.)

Allwood, Abigail C., Malcolm R. Walter, Balz S. Kamber, Craig P. Marshall, and Ian W. Burch. "Stromatolite Reef From the Early Archaean Era of Australia." *Nature* 441, (June 8, 2006): 714–18.

Amthor, Joachim E., John P. Grotzinger, Stefan Schröder, Samuel A. Bowring, Jahandar Ramezani, Mark W. Martin, and Albert Matter. "Extinction of *Cloudina* and *Namacalathus* at the Precambrian-Cambrian Boundary in Oman." *Geology* 31:5 (May 2003): 431–34.

Andres, Miriam S., and R. Pamela Reid. "Growth Morphologies of Modern Marine Stromatolites: A Case Study from Highborne Cay, Bahamas." *Sedimentary Geology* 185 (2006): 319–28.

Angier, Natalie. *The Canon: A Whirligig Tour of the Beautiful Basics of Science.* New York: Houghton Mifflin, 2007.

Anker, Peder. "The Ecological Colonization of Space." *Environmental History* 10:2 (April 2005): 239–68.

Arkin, Adam P. and Daniel A. Fletcher. "Fast, Cheap and Somewhat in Control." *Genome Biology* 7:8 (2006): 1–6.

Arnaud, Celia. "Artificial P450 Enzymes Created." *Chemical and Engineering News* 84:16 (April 17, 2006): 7.

Arnold, Frances H. "Directed Enzyme Evolution." Frances H. Arnold Research Group. http://www.che.caltech.edu/groups/fha/Enzyme/directed.html.

―――. "Design by Directed Evolution." *Accounts of Chemical Research* 31:3 (1998): 125–31.

―――. "Fancy Footwork in the Sequence Space Shuffle." *Nature Biotechnology* 24:3 (March 2006): 328–30.

Arnold, Frances H., and Christopher R. Otey. "Libraries of Optimized Cytochrome P450 Enzymes and the Optimized P450 Enzymes." United States Patent Application 20050059045. March 17, 2005. Accessed via http://www.uspro .gov, January 20, 2008.

Arnold, Frances H. "Unnatural Selection: Molecular Sex for Fun and Profit." *Engineering & Science* 62:1–2 (1999): 40–50.

Asimov, Isaac. *A Short History of Chemistry*. Westport, Conn.: Greenwood Press, 1979.

―――. *The World of Carbon*. New York: Abelard-Schuman, 1958.

Atkins, Peter. *Atkins' Molecules*. 2nd ed. Cambridge: Cambridge University Press, 2003.

―――. *Atoms, Electrons and Change*. New York: Scientific American Library, 1991.

Ausubel, Jesse H. "Decarbonization: The Next 100 Years." Alvin Weinberg lecture presented at Oak Ridge National Laboratory, June 5, 2003.

Ausubel, Jesse. "Energy and Environment: The Light Path." *Energy Systems and Policy* 15:3 (1991):181–88.

―――. "Renewable and Nuclear Heresies." Plenary address to the Canadian Nuclear Association, March 10, 2005.

―――. "Where Is Energy Going?" *Industrial Physicist* 6:1 (February 2000): 16–19.

Ausubel, Jesse H., and Cesare Marchetti. "The Evolution of Transport." *Industrial Physicist* 7:2 (April/May 2001): 20–24.

―――. "Toward Green Mobility: The Evolution of Transport." *European Review* 6:2 (1998): 137–56.

Avise, John C. "The Real Message from Biosphere 2." *Conservation Biology* 8:2 (June 1994): 327–29.

Azam, Farooq. "Microbial Control of Oceanic Carbon Flux: The Plot Thickens." *Science* 280 (May 1, 1998): 694–96.

Azam, Farooq, and Alexandra Z. Worden. "Microbes, Molecules, and Marine Ecosystems" *Science* 303 (March 12, 2004): 1622–24.

Bada, Jeffrey. "How Life Began on Earth: A Status Report." *Earth and Planetary Science Letters* 226 (2004): 1–15.

Bada, Jeffrey L., and Antonio Lazcano. "Prebiotic Soup—Revisiting the Miller Experiment." *Science* 300 (May 2, 2003): 745–46.

Baggott, Jim. *Perfect Symmetry: The Accidental Discovery of Buckminsterfullerene.* New York: Oxford University Press, 1995.

Baker, David, George Church, Jim Collins, Drew Endy, Joseph Jacobson, Jay Keasling, Paul Modrich, Christina Smolke, and Ron Weiss. "Engineering Life: Building a FAB for Biology." *Scientific American,* 294:6. (June 2006): 44–51.

Baker, David F. "Reassessing Carbon Sinks." *Science* 316 (June 22, 2007): 1708–9.

Ball, Philip. "What Is Life? Can We Make It?" *Prospect Magazine* 101 (August 2004): http://www.prospect-magazine.co.uk/article_details.php?id=6205.

Balter, Michael. "Radiocarbon Dating's Final Frontier." *Science* 313 (September 15, 2006): 1560–63.

Bamshad, Michael, and Stephen P. Wooding. "Signatures of Natural Selection in the Human Genome." *Nature Reviews: Genetics* 4 (February 2003): 99–111.

Barrow, John D. *The Constants of Nature: The Numbers that Encode the Deepest Secrets of the Universe.* New York: Vintage, 2002.

Bartlett, P. D., F. H. Westheimer, and G. Buchi. "Robert Burns Woodward, Nobel Prize in Chemistry for 1965." *Science* 150 (October 29, 1965): 585–87.

Barton, Derek H. R. "The Principles of Conformational Analysis." Nobel Lecture, December 11, 1969. In *Nobel Lectures, Chemistry 1963–1970.* Amsterdam: Elsevier Publishing Company, 1972. http://nobelprize.org/nobel_prizes/chemistry/laureates/1969/barton-lecture.pdf. Accessed April 21, 2007.

Barton, Nicholas H., Derek E. G. Briggs, Jonathan A. Eisen, David B. Goldstein, and Nipam H. Patel. *Evolution.* Cold Spring Harbor, New York: Cold Spring Harbor Laboratory Press, 2007.

Becker, Luann, Robert J. Poreda, and Ted E. Bunch. "Fullerenes: An Extraterrestrial Carbon Carrier Phase for Noble Gases." *Proceedings of the National Academy of Sciences* 97:7 (March 28, 2000): 2979–83.

Beerling, David J. *Emerald Planet: How Plants Changed Earth's History.* New York: Oxford University Press, 2007.

Beerling, David J., C. P. Osborne, and W. G. Chaloner. "Evolution of Leaf-Form in Land Plants Linked to Atmospheric CO_2 Decline in the Late Paleozoic Era." *Nature* 410 (March 15, 2001): 352–54.

Beerling, David J., and Robert A. Berner. "Feedbacks and the Coevolution of Plants and Atmospheric CO_2." *Proceedings of the National Academy of Sciences* 102:5 (February 1, 2005): 1302–5.

————. "Impact of a Permo-Carboniferous High O_2 Event on the Terrestrial Carbon Cycle." *Proceedings of the National Academy of Sciences* 97:23 (November 7, 2000): 12428–32.

Beerling, David J., B. H. Lomax, D. L. Royer, G. R. Upchurch Jr., and L. R. Kump. "An Atmospheric pCO_2 Reconstruction Across the Cretaceous-Tertiary Boundary From Leaf Megafossils." *Proceedings of the National Academy of Sciences* 99:12 (June 11, 2002): 7836–40.

Beerling, David, and Dana Royer. "Fossil Plants as Indicators of the Phanerozoic Global Carbon Cycle." *Annual Review of Earth and Planetary Sciences* 30 (2002): 527–56.

Beerling, D. J., and D. L. Royer. "Reading a CO_2 Signal from Fossil Stomata." *New Phytologist* 153 (2002): 387–97.

Behar, Doron M., Saharon Rosset, Jason Blue-Smith, Oleg Balanovsky, Shay Tzur, David Comas, R. John Mitchell, Lluis Quintana-Murci, Chris Tyler-Smith, R. Spencer Wells, and the Genographic Consortium. "The Geonographic Project Public Participation Mitochondrial DNA Database." *PLoS Genetics* 3:6 (June 2007): 1083–95.

Benfrey, Otto Theodore, and Peter J. T. Morris. *Robert Burns Woodward: Architect and Artist in the World of Molecules*. Philadelphia: Chemical Heritage Press, 2001.

Bengtson, Stefan. "Origins and Early Evolution of Predation." *Paleontological Society Papers* 8 (2002): 289–317.

Bengtson, Stefan, and Yue Zhao. "Predatory Borings in Late Precambrian Mineralized Exoskeletons." *Science* 257 (July 17, 1992): 367–69.

Benner, Steven A. "Redesigning Genetics" *Science* 306 (October 22, 2004): 625–26.

Benner, Steven A., and Michael A. Sismour. "Synthetic Biology." *Nature Reviews: Genetics* 6:7 (July 2005): 533–43.

Benton, Michael J., and Francisco J. Ayala. "Dating the Tree of Life." *Science* 300 (June 13, 2003): 1698–1700.

Benton, Michael J., and Richard Twitchett. "How to Kill (Almost) All Life: The End-Permian Extinction Event." *Trends in Ecology and Evolution* 18:7 (July 2003): 358–65.

Benyus, Janine. *Biomimicry: Innovation Inspired By Nature*. New York: Quill, 2002.

Berge, Christine, and Elisabeth Daynes. "Modeling Three-Dimensional Sculptures of Australopithecines (*Australopithecus afarensis*) for the Museum of Natural History of Vienna (Austria): The Post-Cranial Hypothesis." *Comparative Biochemistry and Physiology—Part A: Molecular and Integrative Physiology* 131:1 (December 2001): 145–57.

Berner, Robert. *The Phanerozoic Carbon Cycle.* New York: Oxford University Press, 2004.

Berner, Robert A. "Atmospheric Oxygen over Phanerozoic Time." *Proceedings of the National Academy of Sciences* 96:20 (September 28, 1999): 10955–57.

———. "The Carbon Cycle and CO_2 over Phanerozoic Time: The Role of Land Plants." *Philosophical Transactions of the Royal Society of London B.* 353 (1998): 75–82.

———. "The Long-Term Carbon Cycle, Fossil Fuels and Atmospheric Composition." *Nature* 426 (November 20, 2003): 323–26.

———. "A New Look at the Long-Term Carbon Cycle." *GSA Today* 9:11 (November 1999): 1–2.

———. "The Rise of Plants and Their Effect on Weathering and Atmospheric CO_2." *Science* 276 (April 25, 1997): 544–46.

Bernstein, Jeremy. *Prophet of Energy: Hans Bethe.* New York: Elsevier-Dutton Publishing, 1981.

Bethe, Hans A. "Energy Production in Stars." Nobel Lecture, December 11, 1967. *Nobel Lectures, Physics 1963–1970.* Amsterdam: Elsevier Publishing Company, 1972.

———. "My Life in Astrophysics." *Annual Review of Astronomy and Astrophysics* 41 (2003): 1–14.

Bio-Economic Research Associates (Bio-era). *Genome Synthesis and Design Futures; Implications for the U.S. Economy.* A special Bio-era report, sponsored by the U.S. Department of Energy, February 2007. Cambridge, Mass.

Biederman, Irving, and Edward A. Vessel. "Perceptual Pleasure and the Brain." *American Scientist* 94 (May–June 2006): 247–53.

Black, Edwin. *Internal Combustion: How Corporations and Governments Addicted the World to Oil and Derailed the Alternatives.* New York: St. Martin's Press, 2006.

Bloomfield, Louis A. "Working Knowledge: Catalytic Converter." *Scientific American* 282:2 (February 2000): 108.

Blout, Elkan. "Robert Burns Woodward 1917–1979: A Biographical Memoir." *Biographical Memoirs.* vol. 80. Washington, D.C.: National Academies Press, 2001.

Bodanis, David. $E=mc^2$: *A Biography of the World's Most Famous Equation.* New York: Berkeley Books, 2000.

Borek, Ernest. *The Atoms Within Us.* New York: Columbia University Press, 1980, 4.

Botta, Oliver, and Jeffrey L. Bada. "The Early Earth." In *The Genetic Code and the Origin of Life,* edited by Lluís Ribas de Pouplana. New York and Georgetown, Tex.: Kluwer Academic and Landes Bioscience, 2004.

Bramble, D. M., and D. E. Lieberman. "Endurance Running and the Evolution of *Homo*." *Nature* 432 (November 18, 2004): 345–52.

Braquet, Pierre, ed. *Ginkgolides: Chemistry, Biology, Pharmacology and Clinical Perspectives*. Barcelona: J. R. Prous Science Publishers, 1988.

Brasier, Martin D., Owen R. Green, Andrew P. Jephcoat, Annette K. Kleppe, Martin J. Van Kranendonk, John F. Lindsay, Andrew Steele, and Nathalie V. Grassinea. "Questioning the Evidence for Earth's Oldest Fossils." *Nature* 416 (March 7, 2002): 76–81.

Brasier, Martin D., Nicola McLoughlin, Owen Green, and David Wacey. "A Fresh Look at the Fossil Evidence for Early Archaean Cellular Life." *Philosophical Transactions of the Royal Society B*, 361:1470 (June 29, 2006): 887–902.

Broad, William. "Too Rich a Soil: Scientists Find the Flaw That Undid the Biosphere." *New York Times*, October 5, 1993, C1.

———. "Biosphere Gets Pure Oxygen to Combat Health Woes." *New York Times*, January 26, 1993: C4.

———. "Paradise Lost: Biosphere Retooled as Atmospheric Nightmare." *New York Times*, November 19, 1996: C1.

Brock, William H. *The Chemical Tree: A History of Chemistry*. New York: Norton, 2000.

Brocks, Jochen J. D., Roger E. Love, Andrew H. Summons, Graham A. Knoll, Logan and Stephen A. Bowden. "Biomarker Evidence for Green and Purple Sulphur Bacteria in a Stratified Palaeoproterozoic Sea." *Nature* 437 (October 6, 2005): 866–70.

Brocks, Jochen J., and Ann Pearson. "Building the Biomarker Tree of Life." *Reviews in Mineralogy & Geochemistry* 59 (2005): 233–58.

Brocks, Jochen J., Roger Buick, Roger E. Summons, and Graham A. Logan. "A Reconstruction of Archean Biological Diversity Based on Molecular Fossils From the 2.78 to 2.45 Billion-Year-Old Mount Bruce Supergroup, Hamersley Basin, Western Australia." *Geochimica et Cosmochimica Acta*, 67:22 (2003): 4321–35.

Brocks, Jochen J., Graham A. Logan, Roger Buick, and Roger E. Summons. "Archean Molecular Fossils and the Early Rise of Eukaryotes." *Science*, 285 (August 13, 1999): 1033–36.

Brocks, J. J., and R. E. Summons. "Sedimentary Hydrocarbons, Biomarkers for Early Life." *Treatise on Geochemistry, Volume 8*. William H. Schlesinger. Amsterdam: Elsevier Publishing Company, 2003.

Broecker, Wallace S. *The Role of the Ocean in Climate, Yesterday, Today and Tomorrow*. Palisades, N.Y.: Eldigio Press, 2005.

Brown, James R. "Ancient Horizontal Gene Transfer." *Nature Reviews/Genetics* 4 (February 2003): 121–32.

Brush, Stephen G. "Dynamics of Theory Change in Chemistry: Part 1. The Benzene Problem: 1865–1945." *Studies in History and Philosophy of Science Part A*. 30:1 (1999): 21–79.

Bryson, Bill. *A Short History of Nearly Everything.* New York: Broadway Books, 2003.

Buchanan, Brenda J. *Gunpowder, Explosives and the State: A Technological History.* Burlington, Vt.: Ashgate Publishing Co., 2006.

Bügl, Hans, John P. Danner, Robert J. Molinari, John T. Mulligan, Han-Oh Park, Bas Reichert, David A. Roth, Ralf Wagner, Bruce Budowle, Robert M. Scripp, Jenifer A. L. Smith, Scott J. Steele, George Church, and Drew Endy. "DNA Synthesis and Biological Security." *Nature Biotechnology* 25:6 (June 2007): 627–29.

Roger Buick, David J. Des Marais, and Andrew H. Knoll. "Stable Isotopic Compositions of Carbonates From the Mesoproterozoic Bangemall Group, Northwestern Australia." *Chemical Geology* 123:1–4 (June 20, 1995): 153–71.

Burbidge, E. Margaret. "Watcher of the Skies." *Annual Review of Astronomy and Astrophysics* 32 (1994): 1–36.

Cairns-Smith, A. G. "The First Organisms." *Scientific American,* 253 (June 1985): 90–100.

Cairns-Smith, A. Graham. "Sketches for a Mineral Genetic Material." *Elements* 1 (June 2005): 157–61.

Caldeira, Ken. "What Corals Are Dying to Tell Us About CO_2 and Ocean Acidification." Presented at the Eighth Annual Roger Revelle Commemorative Lecture, Washington, D.C. The National Academies, March 5, 2007.

Calvin, Melvin. *Following the Trail of Light: A Scientific Odyssey.* Washington, D.C.: American Chemical Society, 1992.

Calvin, Melvin. "The Path of Carbon in Photosynthesis." Nobel Lecture, December 11, 1961. *Nobel Lectures, Chemistry 1942–1962.* Amsterdam: Elsevier Publishing Company, 1964.

Calvin, William H. *A Brain for All Seasons: Human Evolution and Climate Change.* Chicago: University of Chicago Press, 2002.

Campbell, Neil A., Jane B. Reece, Lisa A. Urry, Michael L. Cain, Steven A. Wasserman, Peter V. Minorsky, and Robert B. Jackson. *Biology* 8th ed. San Francisco: Pearson Benjamin Cummings, 2008.

Camus, Albert. *The Myth of Sisyphus and Other Essays.* New York: Vintage International, 1983.

Canadell, Josep G., Corinne LeQuere, Michael R. Raupach, Christopher B. Field, Erik T. Buitenhuis, Philippe Ciais, Thomas J. Conway, Nathan P. Gillett, R. A. Houghton, and Gregg Marland. "Contributions to Accelerating Atmospheric

CO$_2$ Growth From Economic Activity, Carbon Intensity, and Efficiency of Natural Sinks." *Proceedings of the National Academy of Sciences* 104:47 (November 20, 2007): 18866–70.

Canfield, Don E., Simon W. Poulton, Guy M. Narbonne. "Late-Neoproterozoic Deep-Ocean Oxygenation and the Rise of Animal Life," *Science* 315 (January 5, 2007): 92–95.

"Carbon Dubbs Is Dead at 81; Expert on Oil." *Chicago Daily Tribune*, August 25, 1962.

Carrier, David. "The Energetic Paradox of Human Running and Hominid Evolution." *Current Anthropology* 25:4 (August–October 1984) 483–95.

Carroll, Sean B. *Endless Forms Most Beautiful: The New Science of Evo Devo and the Making of the Animal Kingdom.* New York: Norton, 2005.

Catling, David C., and Mark W. Claire. "How Earth's Atmosphere Evolved to an Oxic State: A Status Report." *Earth and Planetary Science Letters* 237 (2005): 1–20.

Cavalier-Smith, Thomas, Martin Brasier, and T. Martin Embley. "Introduction: How and When Did Microbes Change the World?" *Philosophical Transactions of the Royal Society B* 361 (2006): 845–50.

Cech, Thomas R. "The Chemistry of Self-Splicing RNA and RNA Enzymes." *Science* 236 (June 19, 1987): 1532–39.

Cech, Thomas R. "Enhanced: The Ribosome Is a Ribozyme." *Science* 289 (August 11, 2000): 878–79.

Cech, Thomas R. "Exploring the New RNA World." Nobelprize.org. December 3, 2004.

Chameides, W. L., and E. M. Perdue. *Biogeochemical Cycles: A Computer-Interactive Study of Earth System Science and Global Change.* New York: Oxford University Press, 1997.

Chan, Leon Y., Sriram Kosuri, and Drew Endy. "Refactoring Bacteriophage T7." *Molecular Systems Biology* 1 (2005): 1–10.

Chartebois, Robert L., and W. Ford Doolittle. "Computing Prokaryotic Gene Ubiquity: Rescuing the Core from Extinction." *Genome Research* 14 (2004): 2469–77.

Chen, Ingfei. "Born to Run." *Discover.* (May 2006): 62–67.

Chen, Irene. "The Emergence of Cells During the Origin of Life." *Science* 314 (December 8, 2006): 1558–59.

Chen, Irene A., Kourosh Salehi-Ashtiani and Jack W. Szostak. "RNA Catalysis in Model Protocell Vesicles." *Journal of the American Chemical Society* 127 (2005): 13213–19.

Chopra, Paras, and Akhil Kammab. "Engineering Life through Synthetic Biology." *In Silico Biology* 6 (2006): 401–10.

Chen, Li-Qun, Cheng-Sen Li, William G. Chaloner, David J. Beerling, Qi-Gao Sun, Margaret E. Collinson, and Peter L. Mitchell. "Assessing the Potential for the Stomatal Characters of Extant and Fossil Ginkgo Leaves to Signal Atmospheric CO_2 Change." *American Journal of Botany* 88:7 (2001): 1309–15.

Christianson, Gale. *Greenhouse: The 200-Year Story of Global Warming.* New York: Walker & Co., 1999.

Chuck, A., T. Tyrrell, I. J. Totterdell, and P. M. Holligan. "The Oceanic Response to Carbon Emissions Over the Next Century: Investigation Using Three Ocean Carbon Cycle Models." *Tellus* 57 B, (2005) 70–86.

Chyba, Christopher F. "Rethinking Earth's Early Atmosphere." *Science* 308 (May 13, 2005): 962–63.

Cody, George D. "Geochemical Connections to Primitive Metabolism." *Elements* 1 (2005): 139–43.

Cohen, Joel E., and David Tilman. "Biosphere 2 and Biodiversity: The Lessons So Far." *Science* 274 (November 15, 1996): 1150–51.

Colby, Joy Hakanson. "Yoko Brings Her Best Wishes." *Detroit News,* April 27, 2000, 1.

Cole, G. A., M. A. Abu-Ali, S. M. Aoudeh, W. J. Carrigan, H. H. Chen, E. L. Colling, W. J. Gwathney, A. A. Al-Hajji, H. I. Halpern, P. J. Jones, S. H. Al-Sharidi, and M. H. Tobey. "Organic Geochemistry of the Paleozoic Petroleum System of Saudi Arabia." *Energy & Fuels* 8 (1994): 1425–42.

Collins, James. "Who's Afraid of Synthetic Biology?" (Unpublished manuscript). E-mail to the author, November 21, 2007.

Corey, E. J. "The Logic of Chemical Synthesis: Multistep Synthesis of Complex Carbogenic Molecules." Nobel Lecture, December 8, 1990. In *Nobel Lectures, Chemistry 1981–1990*, Editor in charge Tore Frängsmyr in collaboration with editor Bo G. Malmström. Singapore: World Scientific Publishing Co., 1992.

Corey, E. J., and Cheng Xue-Min. *The Logic of Chemical Synthesis.* New York: John Wiley & Sons, 1989.

Corey, E. J., Alan K. Long, and Stewart D. Rubenstein. "Computer-Assisted Analysis in Organic Synthesis." *Science* 228 (April 26, 1985): 408–18.

Corey, E. J., Myung-chol Kang, Manoj C. Desai, Arun K. Ghosh, and Ioannis N. Houpis. "Total Synthesis of (±)-Ginkgolide B." *Journal of the American Chemical Society* 110 (1988): 649–51.

Crimmins, Michael T., Jennifer M. Pace, Philippe G. Nantermet, Agnes S. Kim-Meade, James B. Thomas, Scott H. Watterson, and Allan S. Wagman. "The Total Synthesis of (±)-Ginkgolide B." *Journal of the American Chemical Society* 122 (2000): 8453–63.

Crosby, Alfred W. *Children of the Sun: A History of Humanity's Unappeasable Appetite for Energy.* New York: Norton, 2006.

Cuffey, Kurt M., and Francoise Vimeux. "Covariation of Carbon Dioxide and Temperature from the Vostok Ice Core after Deuterium-Excess Correction." *Nature* 412 (August 2, 2001): 523–27.

Curl, Robert F., Richard E. Smalley, Harold W. Kroto, Sean O'Brien, and James R. Heath. "How the News That We Were Not the First to Conceive of Soccer Ball C_{60} Got to Us." *Journal of Molecular Graphics and Modelling* 19:2 (April 2001): 185–86.

Dalton, Alan B., Steve Collins, Edgar Muñoz, Joselito M. Razal, Von Howard Ebron, John P. Ferraris, Jonathan N. Coleman, Bog G. Kim, Ray H. Baughman. "Super-Tough Carbon-Nanotube Fibres." *Nature* 423 (12 June 2003): 703.

Dalton, Rex. "Squaring Up Over Ancient Life." *Nature* 417 (June 20, 2002): 782–84.

Darwin, Charles Robert, 1872. *The Origin of Species by Means of Natural Selection, or the Preservation of Favoured Races in the Struggle for Life.* London: John Murray. 6th edition; with additions and corrections. Transcribed for John van Wyhe 2002; formatting converted by AEL Data 2006. RN1. http://darwin-online.org.uk/content/frameset?itemID=F391&viewtype=text&pageseq=1. Accessed February 8, 2008.

Darwin, Francis, ed. *The Life and Letters of Charles Darwin, vol. III.* London: John Murray, 1887. Scanned for Darwin Online April 2006; transcribed (double key) by AEL Data July 2006. http://darwin-online.org.uk/content/frameset?itemID=F1452.3&viewtype=side&pageseq=1. Accessed February 8, 2008.

Dawkins, Richard. *The Blind Watchmaker: Why the Evidence of Evolution Reveals a Universe Without Design.* New York: Norton, 1986.

———. *The Selfish Gene.* 30th anniversary ed. New York: Oxford University Press, 2006.

Dawkins, R., and J. R. Krebs. "Arms Races Between and Within Species." *Proceedings of the Royal Society of London, Series B, Biological Sciences* 205 (1979): 489–511.

Deamer, David. "Origins of Membrane Structure." In Lynn Margulis, Clifford Matthews, and Aaron Haselton, eds. *Environmental Evolution: Effects of the Origin of Evolution of Life on Planet Earth* 2nd ed. Cambridge, Mass.: MIT Press, 2000, 67–82.

Deamer, David, Jason P. Dworkin, Scott A. Sandford, Max P. Bernstein, and Louis J. Allamandola. "The First Cell Membranes." *Astrobiology* 2:4 (2002): 371–81.

Deckert, Gerard, Patrick V. Warren, Terry Gaasterland, William G. Young, Anna L.

Lenox, David E. Graham, Ross Overbeek, Marjory A. Snead, Martin Keller, Monette Aujay, Robert Huberk, Robert A. Feldman, Jay M. Short, Gary J. Olsen, and Ronald V. Swanson. "The Complete Genome of the Hyperthermophilic Bacterium Aquifex aeolicus." *Nature* 392 (March 26, 1998): 353–58.

Deans, Tara L., Charles R. Cantor, and James J. Collins. "A Tunable Genetic Switch Based on RNAi and Repressor Proteins for Regulating Gene Expression in Mammalian Cells." *Cell* 130 (July 27, 2007): 363–72.

De Duve, Christian. "The Beginnings of Life on Earth." *American Scientist* (September–October 1995): 50–57.

———. "The Birth of Complex Cells." *Scientific American* 274 (April 1996): 50–57.

DeFeudis, Francis V., Vassilios Papadopoulos, and Katy Drieu. "*Ginkgo biloba* Extracts and Cancer: A Research Area in Its Infancy." *Fundamental and Clinical Pharmacology* 18 (August 2003): 405–17.

Deffeyes, Kenneth. *Hubbert's Peak: The Impending World Oil Shortage.* Princeton, N.J.: Princeton University Press, 2001.

Delsuc, Frederic, Henner Brinkmann, and Philippe Herve. "Phylogenomics and the Reconstruction of the Tree of Life." *Nature Reviews Genetics* 6 (May 2005): 361–75.

Del Tredici, Peter. "Ginkgo in America." *Arnoldia* 41:4 (July 1981): 150–61.

———. "Ginkgo and People: A Thousand Years of Interaction." *Arnoldia* 51:2 (1991): 3–15.

———. "The Evolution, Ecology, and Cultivation of *Ginkgo biloba*." In *Ginkgo biloba*, edited by T. van Beek. Amsterdam: Harwood Academic Publishers, 2000.

———. "The Phenology of Sexual Reproduction in Ginkgo biloba: Ecological and Evolutionary Implications." *Botanical Review* 73:4 (2007): 267–78.

———. "Where the Wild Ginkgos Grow." *Arnoldia* 52:4 (1992): 2–11.

Des Marais, David J. "When Did Photosynthesis Emerge on Earth?" *Science* 289 (September 8, 2000): 1703–5.

Des Marais, David J., Harald Strauss, Roger E. Summons, and J. M. Hayes. "Carbon Isotope Evidence for the Stepwise Oxidation of the Proterozoic Environment." *Nature* 359 (October 15, 1992): 605–9.

Desmond, Adrian, and James Moore. *Darwin: The Life of a Tormented Evolutionist.* New York: Norton, 1991.

Diamond, Jared. *Collapse: How Societies Chose to Fail or Succeed.* New York: Viking, 2005.

———. *Guns, Germs and Steel: The Fates of Human Societies.* New York: Norton, 1999.

———. *The Third Chimpanzee: Evolution and the Future of the Human Animal.* New York: HarperPerennial, 1993.

Ditty J. L., S. B.Williams, and S. S. Golden. "A Cyanobacterial Circadian Timing Mechanism." *Annual Review of Genetics* 37 (2003): 513–43.

Dobzhansky, Thodosius. "Nothing in Biology Makes Sense Except in Light of Evolution." *American Biology Teacher* 35 (March 1973): 125–29.

Doolittle, W. Ford. "Uprooting the Tree of Life." *Scientific American* (February 2000): 90–95.

Drury, Stephen. *Stepping Stones: The Making of Our Home World.* Oxford: Oxford University Press, 1999.

Dukes, Jeff. "Burning Buried Sunshine: Human Consumption of Ancient Solar Energy." *Climatic Change* 61:1–2 (November 2003): 31–44.

Durham, Louise. "The Elephant of All Elephants—History and Geology of Saudi Arabia's Ghawar Field." *Energy Bulletin* (January 4, 2005): http://www .energybulletin.net/3889.html. Accessed May 21, 2007.

Dworkin, Jason P., David W. Deamer, Scott A. Sandford, Louis J. Allamandola. "Self-Assembling Amphiphilic Molecules: Synthesis in Simulated Interstellar Precometary Ices." *Proceedings of the National Academy of Sciences* 98:3 (January 30, 2001): 815–19.

Dyson, Freeman J. "Search for Artificial Sources of Infrared Radiation." *Science* 131 (June 3, 1960): 1667–68.

Eherenfreund, Pascale, and Steven B. Charnley. "Organic Molecules in the Interstellar Medium, Comets and Meteorites: A Voyage from Dark Clouds to the Early Earth." *Annual Review of Astronomy and Astrophysics.* 38 (2000): 427–83.

Eigenbrode, Jennifer L., and Katherine H. Freeman. "Late Archean Rise of Aerobic Microbial Ecosystems." *Proceedings of the National Academy of Sciences* 103:43 (October 24, 2006): 15759–64.

Elderfield, Henry. "Carbonate Mysteries." *Science* 296 (May 31 2002): 1618–21.

Eliot, Charles W., ed. *The Harvard Classics: Scientific Papers* vol. 30. New York: P. F. Collier & Son Company, 1910.

Emsley, John. *Molecules at an Exhibition.* Oxford: Oxford University Press, 1998.

ENCODE Project Consortium. "Identification and Analysis of Functional Elements in 1% of the Human Genome by the ENCODE Pilot Project." *Nature* 447 (June 14, 2007): 799–816.

Encyclopedia Britannica online. "Chemical Element Abundances in Earth's Crust." http://search.eb.com/eb/art-915. Accessed February 20, 2008.

Endy, Drew. "Foundations of Synthetic Biology." *Nature* 438 (November 24, 2005): 449–53.

Endy, Drew, and Michael B. Yaffe. "Signal Transduction: Molecular Monogamy." *Nature* 426 (December 11, 2003): 614–15.

Energy Information Administration. "Chapter 4—Natural Gas." International Energy Outlook 2007 (May 2007): http://www.eia.doe.gov/oiaf/ieo/nat_gas .html. Accessed February 8, 2008.

Erickson, Gregory M. "Gigantism and Comparative Life-History Parameters of Tyrannosoid Dinosaurs." *Nature* 430 (August 12, 2004): 772–75.

Ezrin, Myer. *Plastics Failure Guide: Cause and Prevention.* New York: Hanser Publishers, 1996.

———. "Plastics Failure/People Failure." *Plastics World* 55:2 (February 1997): 17.

Falkowski, Paul C., Miriam E. Katz, Andrew H. Knoll, Antonietta Quigg, John A. Raven, Oscar Schofield, F. J. R. Taylor. "The Evolution of Modern Eukaryotic Phytoplankton." *Science* 305 (July 16, 2004) 354–60.

Falkowski, P., R. J. Scholes, E. Boyle, J. Canadell, D. Caneld, J. Elser, N. Gruber, K. Hibbard, P. Högberg, S. Linder, F. T. Mackenzie, B. Moore III, T. Pedersen, Y. Rosenthal, S. Seitzinger, V. Smetacek, W. Steffen. "The Global Carbon Cycle: A Test of Our Knowledge of Earth as a System." *Science*, 290 (October 13, 2000): 291–96.

Feely, Richard A., Christopher L. Sabine, Kitack Lee, Will Berelson, Joanie Kelypas, Victoria J. Fabry, Frank J. Millero. "Impact of Anthropogenic CO_2 on the $CaCO_3$ System in the Oceans." *Science* 305 (July 16, 2004): 362–66.

Fenichell, Stephen. *Plastics: The Making of a Synthetic Century.* New York: HarperCollins, 1996.

Fennel, Katja, Mick Follows, and Paul G. Falkowski. "The Co-Evolution of the Nitrogen, Carbon and Oxygen Cycles in the Proterozoic Ocean." *American Journal of Science*, 305 (June, September, October, 2005): 526–45.

Ferber, Dan. "Microbes Made to Order." *Science* 303 (January 9, 2004): 158–61.

Ferris, James P. "Mineral Catalysis and Prebiotic Synthesis: Montmorillonite-Catalyzed Formation of RNA." *Elements* 1 (June 2005): 145–49.

Ferris, James P. "Montmorillonite-Catalysed Formation of RNA Oligomers: The Possible Role of Catalysis in the Origins of Life." *Philosophical Transactions of the Royal Society of London B* 361 (2006): 1777–86.

Feynman, Richard P. *The Meaning of It All.* New York: Basic Books, 2005.

———. *QED: The Strange Theory of Light and Matter.* Princeton: Princeton University Press, 1988.

———. *Six Easy Pieces: Essentials of Physics Explained by its Most Brilliant Teacher.* New York: Basic Books, 1995.

Field, Christopher B. "Alarming Acceleration in CO_2 Emissions Worldwide—How Do We Respond?" Science Week 2007: Climate Action, conference presentation. Carnegie Institution for Science, October 22, 2007.

Flannery, Tim. *The Weather Makers: How Man Is Changing the Climate and What It Means for Life on Earth.* New York: Grove Press, 2005.

Fleming, James Rodger. *Historical Perspectives on Climate Change.* New York: Oxford University Press, 1998.

Forte, John. "Bio 1A: General Biology at UC–Berkeley, Fall 2006." Webcast.berkeley.http://webcast.berkeley.edu/course_details.php?seriesid=1906978335. Accessed September 21, 2006–November 26, 2006.

Fortey, Richard. *Trilobite: Eyewitness to Evolution.* New York: Knopf, 2000.

Fountain, Henry. "Antique Nanotubes." *New York Times*, October 28, 2006.

Fowler, William A. "Experimental and Theoretical Nuclear Astrophysics: The Quest for the Origin of the Elements." Nobel lecture, December 8, 1983.

William A. Fowler (1911–1995). Interview by John Greenberg and Carol Bugé. May 3, 1983 through May 31, 1984, October 3, 1986. Archives California Institute of Technology, Pasadena, California.

Frank-Kamenetskii, Maxim D. *Unraveling DNA: The Most Important Molecule of Life.* New York: Basic Books, 1997.

Freese, Barbara, *Coal: A Human History.* New York: Penguin Books, 2003.

Friedman, Thomas L. *The World Is Flat: A Brief History of the 21st Century.* New York: Farrar Straus and Giroux, 2005.

Friend, Tim. *The Third Domain: The Untold Story of Archaea and the Future of Biotechnology.* Washington, D.C.: Joseph Henry Press, 2007.

Fujimoto, Takahiro. *Competing to Be Really, Really Good.* Tokyo: International House of Japan, 2007.

Fuller, R. Buckminster. *Operating Manual for Spaceship Earth.* Buckminster Fuller Institute. http://www.bfi.org/?q=node/419. Accessed December 2007.

Fynbo, Hans O. U., Christian A. Diget, Uffe C. Bergmann, Maria J. G. Borge, Joakim Cederkäll, Peter Dendooven, Luis M. Fraile, Serge Franchoo, Valentin N. Fedosseev, Brian R. Fulton, Wenxue Huang, Jussi Huikari, Henrik B. Jeppesen, Ari S. Jokinen, Peter Jones, Björn Jonson, Ulli Köster, Karlheinz Langanke, Mikael Meister, Thomas Nilsson, Göran Nyman, Yolanda Prezado, Karsten Riisager, Sami Rinta-Antila, Olof Tengblad, Manuela Turrion, Youbao Wang, Leonid Weissman, Katarina Wilhelmsen, Juha Äystö, and the ISOLDE Collaboration. "Revised Rates for the Stellar Triple-α Process From Measurement of ^{12}C Nuclear Resonances." *Nature*, 433 (January 13, 2005): 136–39.

Galik, K., B. Senut, M. Pickford, D. Gommery, J. Treil, A. J. Kuperavage, and R. B. Eckhardt. "External and Internal Morphology of the BAR 1002'00 Orrorin tugenensis Femur." *Science* 305 (Sept 3, 2004): 1450–53.

Gamow, G., and M. A. Tuve. "Technical Report." George Washington University School of Engineering. Archives of the Carnegie Institution of Washington. March 28, 1938.

Gaylarde, C., M. Ribas Silva and T. Warscheid. "Microbial Impact on Building Materials: An Overview." *Materials and Structures*, 36 (June 2003): 342–52.

Gell-Mann, Murray. *The Quark and the Jaguar: Adventures in the Simple and the Complex*. New York: W. H. Freeman and Co., 1994.

Gerstein, Mark B., Can Bruce, Joel S. Rozowsky, Deyou Zheng, Jiang Du, Jan O. Korbel, Olof Emanuelsson, Zhengdong D. Zhang, Sherman Weissman, and Michael Snyder. "What is a Gene, post–ENCODE? History and Updated Definition." *Genome Research* 17 (2007) 669–81.

Gertner, Jon. "From 0 to 60 to World Domination." *New York Times Magazine*. February 18, 2007.

Gertz, H. J., and M. Kiefer. "Review About Ginkgo Biloba Special Extract EGb 761 (Ginkgo)." *Current Pharmaceutical Design* 10 (2004): 261–64.

Gewin, Virginia. "Discovery in the Dirt." *Nature* 439 (January 26, 2006): 384–86.

Gilbert, Walter. "Origin of Life: The RNA World." *Nature* 319 (February 20, 1986): 618.

Gingeras, Thomas R. "Origins of Phenotypes: Genes and Transcripts." *Genome Research* 17 (2007): 682–90.

Gluyas, Jon, and Richard Swarbrick. *Petroleum Geoscience*. Oxford U.K.: Blackwell Publishing 2004.

Gold, Paul E., Larry Cahill, and Gary L. Wenk. "The Lowdown on Ginkgo biloba." *Scientific American* 288 (April 2003): 86–91.

Gold, Thomas. "The Deep, Hot Biosphere." *Proceedings of the National Academy of Sciences* 89 (July 1992): 6045–49.

Golden, Susan S., and Sharon R. Canales. "Cyanobacterial Circadian Clocks— Timing Is Everything." *Nature Reviews: Microbiology* 1:3 (December 2003): 191–99.

Goldblatt, Colin, Timothy M. Lenton, and Andrew J. Watson. "Bistability of Atmospheric Oxygen and the Great Oxidation." *Nature* 443 (October 12, 2006): 683–86.

Goldfarb, Michael A., Terrence F. Ciurej, Michael A. Weinstein, and LeRoy W. Metker. "A Method for Soft Body Armor Evaluation: Medical Assessment." Technical Report, July 1973 to June 1974. Edgewood Arsenal, Aberdeen Proving Ground, Maryland (January 1975).

Golubic, Stjepko. "Microbial Landscapes: Abu Dhabi and Shark Bay." In Lynn Margulis, Clifford Matthews, and Aaron Haselton, eds. *Environmental Evolution: Effects of the Origin of Evolution of Life on Planet Earth* 2nd ed. Cambridge, Mass.: MIT Press, 2000, 117–40.

Goodwin, Brian. *How the Leopard Changed Its Spots: The Evolution of Complexity.* Princeton: Princeton University Press, 2001.

Gosnell, Mariana. *Ice: The Nature, the History, and the Uses of an Astonishing Substance.* New York: Knopf, 2005.

Gott, J. Richard III. "Implications of the Copernican Principle for Our Future Prospects." *Nature* 363 (May 27, 1993): 315–19.

Gott, J. Richard III. *Time Travels in Einstein's Universe: The Physical Possibilities of Travel Through Time.* New York: Houghton Mifflin, 2001.

Gould, Stephen Jay. *Wonderful Life: The Burgess Shale and the Nature of History.* New York: Norton, 1989.

Gowdy, John. "Behavioral Economics and Climate Change Policy." (Forthcoming.)

Graham, Linda E., Martha E. Cook, and James S. Busse. "The Origin of Plants: Body Plan Changes Contributing to a Major Evolutionary Radiation." *Proceedings of the National Academy of Sciences* 97:9 (April 25, 2000): 4535–40.

Gray, Harry B. *Chemical Bonds: An Introduction to Atomic and Molecular Structure.* Sausalito, Calif.: University Science Books, 1994.

Gray, Harry B., John D. Simon, and William C. Trogler. *Braving the Elements.* Sausalito, Calif: University Science Books, 1995.

Greally, John M. "Encyclopaedia of Humble DNA." *Nature* 447 (June 14, 2007): 782–83.

Greb, Stephen F., William A. DiMichele, and Robert A. Gastaldo. "Evolution and Importance of Wetlands in Earth History." Geological Society of America Special Paper No. 399, 2006.

Greb, Stephen F., W. M. Andrews, C. F. Eble, W. DiMichele, C. B. Cecil, and J. C. Hower. "Desmoinesian Coal Beds of the Eastern Interior and Surrounding Basins: The Largest Tropical Peat Mires in Earth History." Geological Society of America Special Paper No. 370, 2003.

Gregory, Jane. *Fred Hoyle's Universe.* Oxford: Oxford University Press, 2005.

Grinspoon, David. *Lonely Planets: The Natural Philosophy of Alien Life.* New York: Ecco, 2004.

Grotzinger, John P., Samuel A. Bowing, Beverly Z. Saylor, Alan J. Kaufman. "Biostratigraphic and Geochronologic Constraints on Early Animal Evolution. *Science* 270 (October 27, 1995): 598–604.

Haberl, Helmut, K. Heinz Erb, Fridolin Krausmann, Veronika Gaube, Alberte Bondeau, Christoph Plutzar, Simone Gingrich, Wolfgang Lucht, and Marina Fischer-Kowalski. "Quantifying and Mapping the Human Appropriation of Net Primary Production in Earth's Terrestrial Ecosystems." *Proceedings of the National Academy of Sciences* 104:31 (July 31, 2007): 12942–47.

Han, Tsu-Ming, and Bruce Runnegar. "Megascopic Eukaryotic Algae from the

2.1-Billion-Year-Old Negaunee Iron Formation, Michigan." *Science* 257 (July 10, 1992): 232–35.

Hanczyc, Martin M., Sheref S. Mansy, and Jack W. Szostak. "Mineral Surface Directed Membrane Assembly." *Origin of Life and Evolution of the Biosphere.*

Handa, Mariko. "*Ginkgo biloba* in Japan." *Arnoldia* 60:4 (2000): 26–33.

Hansen, James, Makiko Sato, Pushker Kharecha, Gary Russell, David W. Lea, and Mark Siddall. "Climate Change and Trace Gases." *Philosophical Transactions of the Royal Society A* 365 (2007): 1925–54.

Hansen, James, Makiko Sato, Reto Ruedy, Ken Lo, David W. Lea, and Martin Medina-Elizade. "Global Temperature Change." *Proceedings of the National Academy of Sciences* 103:39 (September 26, 2006): 14288–93.

Hansen, James. "The Threat to the Planet." *New York Review of Books*, July 13, 2006, 12–16.

Hansen J., M. Sato, R. Ruedy, P. Kharecha, A. Lacis, R. L. Miller, L. Nazarenko, K. Lo, G. A. Schmidt, G. Russell, I. Aleinov, S. Bauer, E. Baum, B. Cairns, V. Canuto, M. Chandler, Y. Cheng, A. Cohen, A. Del Genio, G. Faluvegi, E. Fleming, A. Friend, T. Hall, C. Jackman, J. Jonas, M. Kelley, N. Y. Kiang, D. Koch, G. Labow, J. Lerner, S. Menon, T. Novakov, V. Oinas, Ja. Perlwitz, Ju. Perlwitz, D. Rind, A. Romanou, R. Schmunk, D. Shindell, P. Stone, D. Streets, S. Sun, N. Tausnev, D. Thresher, N. Unger, M. Yao, and S. Zhang. "Dangerous Human-Made Interference With Climate: A GISS modelE study." *Atmospheric Chemistry and Physics* 7 (2007): 2287–2312.

Hansen, James. "Declaration of James E. Hansen." *Green Mountain Chrysler-Plymouth-Dodge-Jeep, et al, Plaintiffs; Association of International Automobile Manufacturers, Plaintiff, v. Thomas W. Torti, Secretary of the Vermont Agency of Natural Resources, et al., Defendants.* United States District Court for the District of Vermont. Case Nos. 2:05-CV-302, and 2:05-CV-304 (consolidated).

Haqq-Misra, Jacob D., Shawn D. Domagal-Goldman, Patrick J. Kasting, and James F. Kasting. "A Revised, Hazy Methane Greenhouse for the Archean Earth." (Forthcoming.)

Harris, Robert, and Jeremy Paxman. *A Higher Form of Killing: The Secret History of Chemical and Biological Wafare.* New York: Random House, 2002.

Hartmann, William K. *Moons and Planets: Fourth Edition.* Belmont, Calif.: Wadsworth Publishing Co., 1999.

Haskin, L. A. "Water and Cheese from the Lunar Desert: Abundances and Accessibility of H, C, and N on the Moon." In W. W. Mendell. ed. *Second Conferences on Lunar Bases and Space Activities in the 21st Century*, Vol. 2. NASA Conferences Publication 3166. Houston: NASA, 1992.

Hay, W. W. "Tectonics and Climate." *Geologische Rundschau* 85:3 (1996): 409–37.

Hayes, John M. "The Pathway of Carbon in Nature." *Science* 312 (June 16, 2006): 1605–1606.

Hayes, John M., and Jacob R. Waldbauer. "The Carbon Cycle and Associated Redox Processes Through Time." *Philosophical Transactions of the Royal Society B* 361 (June 29, 2006): 931–50.

Hazen, Robert. *The Diamond Makers*. Cambridge: Cambridge University Press, 1999.

———. *Genesis: The Scientific Quest for Life's Origins*. Washington, D.C.: Joseph Henry Press, 2005.

———. "Mineral Surfaces and the Prebiotic Selection and Organization of Biomolecules." *American Mineralogist* 91 (2006): 1715–29.

———. "Genesis: Rocks, Minerals, and the Geochemical Origin of Life." *Elements* 1 (2005): 135–37.

Hazen, Robert M., Patrick L. Griffin, James M. Carothers, and Jack W. Szostak. "Functional Information and the Emergence of Biocomplexity." *Proceedings of the National Academy of Sciences* 104 (May 15, 2007): 8574–81.

Hazen, Robert M., and David S. Sholl. "Chiral Selection on Inorganic Crystalline Surfaces." *Nature Materials* 2 (June 2003): 367–74.

Hazen, Robert M. "Why Should You Be Scientifically Literate?" ActionBioscience.org, http://www.actionbioscience.org/newfrontiers/hazen.html. Accessed March 14, 2007.

Heinemann, Mattias, and Sven Panke. "Synthetic Biology—Putting Engineering into Biology." *Bioinformatics* 22:22 (2006): 2790–99.

Heinrich, Bernd. *Racing the Antelope: What Animals Can Teach Us About Running and Life*. New York: Cliff Street Books, 2001.

Helfand, David. Frontiers of Science: *Scientific Habits of Mind Multimedia Study Environment*. Accessed from http://ccnmtl.columbia.edu/projects/mmt/frontiers, on August 27, 2007.

Helge, Jørn W., Bente Stallknecht, Erik A. Richter, Henrik Galbo, and Bente Kiens. "Muscle Metabolism During Graded Quadriceps Exercise in Man." *Journal of Physiology* 581:3 (2007): 1247–58.

Henning, T., and F. Salama. "Carbon in the Universe." *Science* 282, (December 18, 1998): 2204–2210.

Herbst, Eric. "The Chemistry of Interstellar Space." *Chemical Society Review* 30 (2001): 168–76.

———. "Chemistry of Star-Forming Regions." *Journal of Physical Chemistry A* 109:18 (May 12, 2005): 4017–25.

Herz, Werner. *The Shape of Carbon Compounds*. New York: W. A. Benjamin, 1963.

Hillier, Victor, and Peter Coombes. *Hillier's Fundamentals of Motor Vehicle Technology.* London: Nelson Thames, 2004.

Hoelzer, G. A., E. Smith, and J. W. Pepper. "On the Logical Relationship Between Natural Selection and Self-Organization." *Journal of Evolutionary Biology* 19:6 (2006): 1785–94.

Hofmann, H. J. "Precambrian Microflora, Belcher Islands, Canada: Significance and Systematics. *Journal of Paleontology* 50:6 (November 1976) 1040–73.

Hoffman, Paul F., and Daniel P. Schrag. "The Snowball Earth Hypothesis: Testing the Limits of Global Change." *Terra Nova* 14:3 (2002) 129–55.

Hoffmann, Roald. *The Same and Not the Same.* New York: Columbia University Press, 1995.

Hoffmann, Roald. "How Should Chemists Think?" *Scientific American* (February 1993): 66–73.

———. "Molecular Beauty III: As Rich as Need Be." *American Scientist* 77:2 (March–April 1989): 177–78.

———. "Unstable." *American Scientist* 75 (November–December 1987): 619–21.

Holland, Heinrich D. "The Oxygenation of the Atmosphere and Oceans." *Philosophical Transcripts Royal Society of London B* 361:1470 (June 29, 2006): 903–15.

Hori, Shihomi, and Teruaitsu Hori. "A Cultural History of *Ginkgo biloba* in Japan and the Generic Name Ginkgo." In van Beek, Teris, ed. *Ginkgo Biloba.* 386–93.

Houghton, R. A. "Balancing the Global Carbon Budget." *Annual Review of Earth and Planetary Sciences* 35 (2007): 313–47.

Howell, F. Clark. *Early Man.* In collaboration with the editors of Time-Life Books. *Life Nature Library.* New York: Time-Life Books, 1970.

"How to Make Chlorophyll." *Time,* July 18, 1960. Accessed from http://www.time .com/time/magazine/article/0,9171,869621,00.html

Hoyle, Fred. *Home Is Where the Wind Blows: Chapters From a Cosmologist's Life.* Mill Valley, Calif.: University Science Books, 1994.

Hoyle, Fred. "The Universe: Past and Present Reflections." *Annual Review of Astronomy and Astrophysics* 20 (1982): 1–36.

Hsu, Kenneth J., Hedi Oberhänsli, J. Y. Gao, Sun Shu, Chen Haihong, and Urs Krähenbühl. " 'Strangelove Ocean' Before the Cambrian Explosion" *Nature* 316, (August 29, 1985): 809–11.

Hua, Hong, Brian R. Pratt, and Lu-Yi Zhang. "Borings in Cloudina Shells: Complex Predator-Prey Dynamics in the Terminal Proterozoic." *Palaios* 18 (2003): 454–59.

"Huge 2.5 inch Stromatolite Sphere 2,200 Million Years." eBay auction, item number 130051673910. Accessed December 3, 2006.

Hughes, Randall A., Michael P. Robertson, Andrew D. Ellington, and Matthew Levy. "The Importance of Prebiotic Chemistry in the RNA World." *Current Opinion in Chemical Biology* 8 (2004): 629–33.

Huntley, John Warren, and Michal Kowalewski. "Strong Coupling of Predation Intensity and Diversity in the Phanerozoic Fossil Record. *Proceedings of the National Academy of Sciences.* 104:38 (September 18, 2007): 15006–10.

Huxley, T. H. "On a Piece of Chalk." In *Lay Sermons, Addresses and Reviews.* New York: D. Appleton and Company, 1903.

Intergovernmental Panel on Climate Change: *Climate Change 2007: The Physical Science Basis. Contribution of Working Group I to the Fourth Assessment Report of the Intergovernmental Panel on Climate Change.* Edited by S. D. Solomon Qin, M. Manning, Z. Chen, M. Marquis, K. B. Averyt, M. Tignor, and H. L. Miller. Cambridge, U.K. and New York: Cambridge University Press, 2007.

———. *Climate Change 2007: Impacts, Adaptation and Vulnerability. Contribution of Working Group II to the Fourth Assessment Report of the Intergovernmental Panel on Climate Change.* Edited by M. L. Parry, O. F. Canziani, J. P. Palutikov, P. J. van der Linden, and C. E. Hanson. Cambridge, U.K.: Cambridge University Press, 2007.

———. *Climate Change 2007: Mitigation. Contribution of Working Group III to the Fourth Assessment Report of the Intergovernmental Panel on Climate Change.* Edited by B. Metz, O. R. Davidson, P. R. Bosch, R. Dave, and L. A. Meyer. Cambridge, U.K. and New York: Cambridge University Press, 2007.

Isaacs, Farren J., David J. Dwyer, and James J. Collins. "RNA Synthetic Biology." *Nature Biotechnology* 24:5 (May 2006): 545–54.

"ITIS Standard Report Page: Ginkgo biloba," Taxonomic Serial No.: 183269. Accessed from http://www.itis.gov/servlet/SingleRpt/SingleRpt?search_topic= TSN&search_value=183269 on February 8, 2008.

Isaacson, Walter. *Einstein: His Life and Universe.* New York: Simon and Schuster, 2007.

Ishikawa, Eishei, and David L. Swain, trans. "The Committee for the Compilation of Materials on Damage Caused by the Atomic Bombs in Hiroshima and Nagasaki." In *Hiroshima and Nagasaki: The Physical, Medical and Social Effects of the Atomic Bombings,* 87. New York: Basic Books, 1981.

Johnson R. D., and C. Holbrow, eds. *Space Settlements: A Design Study.* NASA, SP-413. Scientific and Technical (1977): http://www.nss.org/settlement/ nasa/75SummerStudy/Design.html.

Jones, Dan. "Personal Effects." *Nature* 438 (November 3, 2005): 14–16.

Joyce, Gerald F. "The Antiquity of RNA-Based Evolution." *Nature* 418 (July 11, 2002): 214–21.

Joyce, Gerald F. "Directed Molecular Evolution." *Scientific American* (December 1992): 90–97.

Kamminga, Harmke. "The Kekulé Riddle." *Lancet* 341 (June 5, 1993): 1463.

Karol, Kenneth G., Richard M. McCourt, Matthew T. Cimino, and Charles F. Delwiche. "The Closest Living Relatives of Land Plants." *Science* 294 (December 14, 2001): 2351–53.

Kasting, James F. "The Rise of Atmospheric Oxygen." *Science* 293 (August 3, 2001): 819–20.

———. "The Carbon Cycle, Climate, and the Long-Term Effects of Fossil Fuel Burning." *Consequences* 4:1 (1998): 15–27.

Kauffman, Stuart. *Investigations.* New York: Oxford University Press, 2000.

Kaufman, Alan J. "The Calibration of Ediacaran Time." *Science* 308, (April 1, 2005): 59–60.

Ke, Yuehai, Bing Su, Xiufeng Song, Daru Lu, Lifeng Chen, Hongyu Li, Chunjian Qi, Sangkot Marzuki, Ranjan Deka, Peter Underhill, Chunjie Xiao, Mark Shriver, Jeff Lell, Douglas Wallace, R. Spencer Wells, Mark Seielstad, Peter Oefner, Dingliang Zhu, Jianzhong Jin, Wei Huang, Ranajit Chakraborty, Zhu Chen, Li Jin. "African Origin of Modern Humans in East Asia: A Tale of 12,000 Y Chromosomes." *Science* 292 (May 11, 2001): 1151–53.

Keeling, Charles D. "A Brief History of Atmospheric CO_2 Measurements and Their Impact on Thoughts about Environmental Change." Asahi Glass Blue Planet Prize lecture, 1993. http://www.af-info.or.jp/eng/honor/bppcl_e/e1993keeling.txt. Accessed June 5, 2007.

———. "Rewards and Penalties of Monitoring the Earth." *Annual Review of Energy and the Environment* 23 (1998): 25–82.

Kelly, Jack. *Gunpowder.* New York: Basic Books, 2004.

Kelly, Kevin. *Out of Control: The New Biology of Machines, Social Systems and the Economic World.* Cambridge, Mass.: Perseus Books, 1994.

Kenrick, Paul, and Peter R. Crane. "The Origin and Early Evolution of Plants on Land." *Nature* 389 (September 4, 1997): 33–39.

Kerr, Richard. "A Shot of Oxygen to Unleash the Evolution of Animals," *Science* 314, (December 8, 2006): 1529.

Kiehl, J. T., and Kevin E. Trenberth. "Earth's Annual Global Mean Energy Budget." *Bulletin of the American Meteorological Society* 78:2 (February 1997): 197–208.

Kirby, Richard Shelton, Sidney Withington, Arthur Burr Darling, and Frederick Gridley Kilgour. *Engineering in History.* Mineola, New York: Dover, 1990.

Kirsch, David A. *The Electric Vehicle and the Burden of History.* New Brunswick, N.J.: Rutgers University Press, 2000.

Kirschner, Marc W., and John C. Gerhart. *The Plausibility of Life*. New Haven, Conn.: Yale University Press, 2005.

Kirschvink, Joseph. "Red Earth, White Earth, Green Earth, Black Earth." *Engineering & Science* 4 (2005): 10–20.

Kirschvink, J. L., and J. W. Hagadorn. "A Grand Unified Theory of Biomineralization." In *The Biomineralization of Nano- and Micro-Structures*, edited by E. Bäuerlein. Weinheim, Germany: Wiley-VCH Verlag GmbH, 2000, 139–50.

Kirschvink, Joseph L., and Robert E. Kopp. "Arguments for the Late Evolution of Oxygenic Photosynthesis at 2.3 Ga: A Trigger for the Paleoproterozoic Snowball Earth." *Geophysical Research Abstracts* 7 (2005): 11197.

Kittler, Ralf, Manfred Kayser, and Mark Stoneking. "Molecular Evolution of *Pediculus humanus* and the Origin of Clothing." *Current Biology* 13:16 (August 19, 2003): 1414–17.

Klein, Richard G., and Blake Edgar. *The Dawn of Culture*. New York: John Wiley & Sons Inc., 2002.

Klemme, H. D., and G. F. Ulmishek. "Effective Petroleum Source Rocks of the World: Stratigraphic Distribution and Controlling Depositional Factors. Search and Discovery Article #30003 (1999) http://searchanddiscovery. com/documents/Animator/Klemme2.htm. Accessed January 2008.

Klemperer, William. "The Chemistry of Interstellar Space." Royal Institution Discourse. Presentation at The Royal Institution, London, 1995. Accessed from http://vega.org.uk/video/programme/64 in May 2006.

Knight, Jonathan. "No Dope." *Nature* 426 (November13, 2003): 114–15.

Knight, Thomas F. "Engineering Novel Life." *Molecular Systems Biology* 1 (2005): 1.

Knoll, Andrew. *Life on a Young Planet: The First Three Billion Years of Evolution on Earth*. Princeton: Princeton University Press, 2003.

Knoll, Andrew H. The Early Evolution of Eukaryotes: A Geological Perspective." *Science*, 256, (May 1, 1992): 622–27.

———. "A New Molecular Window on Early Life." *Science* 285 (August 13, 1999): 1025–26.

———. "The Geological Consequences of Evolution." *Geobiology* 1 (2003): 3–14.

———. "Biomineralization and Evolutionary History." *Biomineralization* 54 (2003): 329–56.

Knoll, Andrew H., and Sean B. Carroll. "Early Animal Evolution: Emerging Views from Comparative Biology and Geology." *Science* 284 (June 25, 1999): 2129–37.

Knoll, Andrew H., Malcolm R. Walter, Guy M. Narbonne, and Nicholas Christie-Blick. "A New Period for the Geologic Time Scale." *Science* 305 (July 30, 2004): 621–22.

Kolbert, Elizabeth. *Field Notes From a Catastrophe*. New York: Bloomsbury USA, 2006.

———. "The Darkening Sea." *New Yorker*, (November 20, 2006), 67–75.

Kopp, Robert E., Joseph L. Kirschvink, Isaac A. Hilburn, and Cody Z. Nash. "The Paleoproterozoic Snowball Earth: A Climate Disaster Triggered by the Evolution of Oxygenic Photosynthesis." *Proceedings of the National Academy of Sciences* 102:32 (August 9, 2005): 11131–36.

Korotev, Randy L. "In Memoriam: Larry Haskin (1934–2005)." *Geochemical News* 123 (April 2005): 8–9.

Krauss, Lawrence M. *Atom: A Single Oxygen Atom's Odyssey from the Big Bang to Life on Earth . . . and Beyond*. New York: Back Bay Books, 2002.

Krebs, Hans. "The Citric Acid Cycle." Nobel Lecture, December 11, 1953. In *Nobel Lectures, Physiology or Medicine 1942–1962*, Amsterdam: Elsevier Publishing Company, 1964.

Kroto, Harold W. "C60: Buckminsterfullerene, The Celestial Sphere that Fell to Earth." *Angewandte Chemie International Edition in English* 31:2 (February 1992): 111–29.

Krugman, Paul. "What Economists Can Learn from Evolutionary Theorists." Address to the European Association for Evolutionary Political Economy, November 1996. http://www.mit.edu/~krugman/evolute.html. Accessed November 15, 2006.

Kuhn, Thomas S. *The Structure of Scientific Revolutions* 3rd ed. Chicago: University of Chicago Press, 1996.

Kump, Lee R. "Reducing Uncertainty about Carbon Dioxide as a Climate Driver." *Nature* 419 (September 12, 2002): 188–90.

Kump, Lee R., James F. Kasting, and Robert G. Crane. *The Earth System*. Upper Saddle River, N.J.: Prentice Hall, 2004.

Kurzweil, Ray, and Bill Joy. "Recipe for Destruction." *New York Times* (October 17, 2005): 19.

Kwolek, Stephanie L. Interviewed by Bernadette Bensaude-Vincent at Wilmington, Delaware, March 21, 1998. Philadelphia: Chemical Heritage Foundation, Oral History Transcript #0168.

Kwolek, Stephanie. "Stephanie Kwolek Innovative Lives Presentation." March 25, 1996. Archives Center, National Museum of American History.

LaBarbera, Michael. "Why Wheels Won't Go." *American Naturalist* 121:3 (March 1983): 395–408.

Landa, Edward R. "Oink if You Love Coal." *Geotimes* (April 2006): 60.

Lane, Nick. *Oxygen*. Oxford: Oxford University Press, 2002.

———. *Power, Sex, Suicide*. Oxford: Oxford University Press, 2005.

Langer, G., M. Geisen, U. Riebesell, J. Kläs, S. Krug, K. H. Baumann, and J. Young.

"The Response of *Calcidiscus leptoporus* and *Coccolithus pelagicus* to Changing Carbonate Chemistry of Seawater." *Geophysical Research Abstracts*, 8 (2006): 05161.

Larsson, B., R. Liseau, L. Pagani, P. Bergman, P. Bernath, N. Biver, J. H. Black, R. S. Booth, V. Buat, J. Crovisier, C. L. Curry, M. Dahlgren, P. J. Encrenaz, E. Falgarone, P. A. Feldman, M. Fich, H. G. Florén, M. Fredrixon, U. Frisk, G. F. Gahm, M. Gerin, M. Hagström, J. Harju, T. Hasegawa, Å. Hjalmarson, L. E. B. Johansson, K. Justtanont, A. Klotz, E. Kyrölä, S. Kwok, A. Lecacheux, T. Liljeström, E. J. Llewellyn, S. Lundin, G. Mégie, G. F. Mitchell, D. Murtagh, L. H. Nordh, L.-Å. Nyman, M. Olberg, A. O. H. Olofsson, G. Olofsson, H. Olofsson, G. Persson, R. Plume, H. Rickman, I. Ristorcelli, G. Rydbeck, A. A. Sandqvist, F. V. Schéele, G. Serra, S. Torchinsky, N. F. Tothill, K. Volk, T. Wiklind, C. D. Wilson, A. Winnberg, and G. Witt. "Molecular Oxygen in the ρ Ophiuchi Cloud." *Astronomy and Astrophysics*, 466:3 (May 2007): 999–1003.

Laurence, William L. "Endless Duel of Atoms Declared Sources of Fuel in Furnace of Sun." *New York Times*, December 18, 1938, 1.

Lazcano, Antonio. "The Origins of Life: Have Too Many Cooks Spoiled the Prebiotic Soup?" *Natural History* (February 2006): 36–41.

Lazcano, Antonio, and Stanley L. Miller. "The Origin and Early Evolution Review of Life: Prebiotic Chemistry, the Pre-RNA World, and Time." *Cell* 85 (June 14, 1996): 793–98.

Le Couteur, Penny, and Jay Burreson *Napoleon's Buttons: 17 Molecules That Changed History*. New York: Tarcher Penguin, 2003.

Lemonick, Michael D. "Cosmic Fingerprint." *Time*, February 24, 2003, 45.

———. "Let There Be Light." *Time*, September 4, 2006.

Levi, Primo. *The Periodic Table*. New York: Schoken Books, 1984.

Levskaya, Anselm, Aaron A. Chevalier, Jeffrey J. Tabor, Zachary Booth Simpson, Laura A. Lavery, Matthew Levy, Eric A. Davidson, Alexander Scouras, Andrew D. Ellington, Edward M. Marcotte, and Christopher A. Voigt. "Engineering *Escherichia coli* to See Light." *Nature* 438 (November 24, 2005): 441–42.

Lewis, Louise A., and Richard M. McCourt. "Green Algae and the Origin of Land Plants." *American Journal of Botany* 91:10 (October 2004): 1535–56.

Lin, Li-Hung, Pei-Ling Wang, Douglas Rumble, Johanna Lippmann-Pipke, Erik Boice, Lisa M. Pratt, Barbara Sherwood Lollar, Eoin L. Brodie, Terry C. Hazen, Gary L. Andersen, Todd Z. DeSantis, Duane P. Moser, Dave Kershaw, and T. C. Onstott. "Long-Term Sustainability of a High-Energy, Low-Diversity Crustal Biome." *Science* 314 (2006): 479–82.

Line, Martin A. "The Enigma of the Origin of Life and Its Timing." *Microbiology* 148 (2002): 21–27.

Longuski, Jim. *The Seven Secrets of How to Think Like Rocket Scientist*. New York: Copernicus Books, 2007.

Lovelock, James. *The Revenge of Gaia*. New York: Basic Books, 2006.

Lovelock, James. "Travels With an Electron Capture Detector." Asahi Glass Blue Planet Prize lecture, 1997. Accessed from http://www.af-info.or.jp/eng/honor/97lect-e.pdf on June 11, 2007.

Lundquist, Stig, ed. *Physics 1971–1980*. Singapore: World Scientific Publishing Co., 1992.

MacMenamin, D., and M. MacMenamin. *The Emergence of Animals: The Cambrian Breakthrough*. New York: Columbia University Press, 1990, 95.

Madsen, Eugene L. "Identifying Microorganisms Responsible for Ecologically Significant Biogeochemical Processes." *Nature Reviews Microbiology* 3 (May 2005): 439–46.

Magat, E. E. "Fibres from Extended Chain Aromatic Polyamides." *Philosophical Transactions of the Royal Society of London A* 294 (1980): 463–72.

Major, Randolph T. "The Ginkgo, the Most Ancient Living Tree." *Science* 157 (September 15, 1967): 1270–73.

Manning, Phillip. *Atoms, Molecules, and Compounds*. New York: Chelsea House Publishers, 2008.

Mantle, Jonathan. *Car Wars*. New York: Little, Brown, 1995.

Marchetti, Cesare. "Nuclear Plants and Nuclear Niches." *Nuclear Science and Engineering* 90 (1985): 521–26.

Marchetti, Cesare. "On Decarbonization: Historically and Perspectively." Interim Report, accessed from http://www.iiasa.ac.at/Admin/PUB/Documents/IR-05-005.pdf in May 2007.

Margulis, Lynn, Clifford Matthews, and Aaron Haseltine, eds. *Environmental Evolution: Effects of the Origin and Evolution of Life on Earth*. Cambridge, Mass. MIT Press, 2000.

Margulis, Lynn. "Symbiosis and the Origin of Protists." In *Environmental Evolution: Effects of the Origin and Evolution of Life on Earth*, 141–70.

Marin Frédéric, Mark Smith, Yeishin Isa, Gerard Muyzer, and Peter Westbroek. "Skeletal Matrices, Muci and the Origin of Invertebrate Calcification." *Proceedings of the National Academy of Sciences*. 93:4 (Feb 20, 1996): 1554–59.

Martin, William, Meike Hoffmeister, Carmen Rotte, and Katrin Henze. "An Overview of Endosymbiotic Models for the Origins of Eukaryotes, Their ATP-Producing Organelles (Mitochondria and Hydrogenosomes), and Their Heterotrophic Lifestyle." *Biological Chemistry* 382:11 (November 2001): 1521–39.

Matsumoto, Kazuho, Takeshi Ohta, Michiya Irasawat, and Tsutomu Nakamurat.

"Climate Change and Extension of the *Ginkgo biloba* L. Growing Season in Japan." *Global Change Biology* 9 (2003): 1634–42.

McKay, Mary Fae, David S. McKay, and Michael B. Duke. "Space Resources: Overview." Lyndon B. Johnson Space Center, Houston, 1992 NASA SP-509 overview.

McManus, Jerry F. "A Great Grand-Daddy of Ice Cores." *Nature* 429 (June 10, 2004): 611–12.

McMurry, John. *Organic Chemistry.* 5th ed. Pacific Grove, Calif.: Brooks/Cole, 2000.

McPhee, John. *The Control of Nature.* New York: Farrar, Straus, and Giroux, 1990.

———. "Season on the Chalk." *New Yorker,* March 12, 2007.

Meert, Joseph G., and Trond H. Torsvik. "The Making and Unmaking of a Supercontinent: Rodinia Revisited." *Tectonophysics* 375 (November 2003): 261–68.

Melezhik, Viktor A. "Multiple Causes of Earth's Earliest Glaciation." *Terra Nova* 18 (2006) 130–37.

Michel-Zaitsu, Wolfgang. "On Engelbert Kaempfer's Ginkgo." Accessed from http://www.flc.kyushu-u.ac.jp/~michel/serv/ek/amoenitates/ginkgo/ginkgo.html on February 12, 2008.

Mitchell, Peter. "David Keilin's Respiratory Chain Concept and Its Chemiosmotic Consequences." Nobel Lecture, December 8, 1978. In *Nobel Lectures, Chemistry 1971–1980.* Editor in charge Tore Frängsmyr in collaboration with editor Sture Forsén. Singapore: World Scientific Publishing Co., 1993

Mitton, Simon. *Conflict in the Cosmos.* Washington, D.C.: Joseph Henry Press, 2005.

Mokyr, Joel. *The Gifts of Athena: Historical Origins of the Knowledge Economy.* Princeton: Princeton University Press, 2002.

———. "Useful Knowledge as an Evolving System: The View from Economic History." Preliminary and incomplete. Revised, February 11, 2002. Paper presented at the conference on The Economy as an Evolving System. Santa Fe, N.M., November 16–18, 2001.

———. "Natural History and Economic History: Is Technological Change an Evolutionary Process?" Draft lecture. April 2000. Accessed at http://faculty.weas.northwestern.edu/~jmokyr/jerusalem.pdf on May 25, 2007.

Montañez, Isabel P., Neil J. Tabor, Deb Niemeier, William A. DiMichele, Tracy D. Frank, Christopher R. Fielding, John L. Isbell, Lauren P. Birgenheier, and Michael C. Rygel. "CO_2-Forced Climate and Vegetation Instability During Late Paleozoic Deglaciation." *Science* 315 (January 5, 2007): 87–91.

Montgomery, Regina R., ed., "NIST SRM 3246 *Ginkgo biloba* Supplement Standard Reference Materials." *SRM Spotlight* (February 2007): 1–11.

Moore, Gordon. "Cramming More Components Onto Integrated Circuits." *Electronics*, 38:8 (April 19, 1965): 1–4. ftp://download.intel.com/museum/Moores_Law/Articles-Press_Releases/Gordon_Moore_1965_Article.pdf.

Morowitz, Harold J. *The Emergence of Everything.* New York: Oxford University Press, 2002.

———. *The Wine of Life and Other Essays on Societies, Energy and Living Things.* New York: St. Martin's Press, 1979.

Morowitz, Harold, and Eric Smith. "Energy Flow and the Organization of Life." *Complexity* 13:1 (October 8, 2007): 51–59.

Morowitz, Harold J., Jennifer D. Kostelnik, Jeremy Yang, and George D. Cody. "The Origin of Intermediary Metabolism." *Proceedings of the National Academy of Sciences* 97:14 (July 5, 2000): 7704–08.

Moskin, Julia. "Creamy, Healthier Ice Cream? What's the Catch?" *New York Times* (July 26, 2006): Fl.

Nakanishi, Koji. *A Wandering Natural Products Chemist.* Washington, D.C.: American Chemical Society, 1991.

———. "Terpene Trilactones from *Ginkgo biloba*: From Ancient Times to the 21st Century." *Bioorganic and Medicinal Chemistry.* 13 (2005): 4987–5000.

National Academy of Sciences. *Teaching About Evolution and the Nature of Science.* Working Group on Teaching Evolution, 0-309-53221-3, 1998.

National Research Council. *Direct and Indirect Human Contributions to Terrestrial Carbon Fluxes: A Workshop Summary.* Rob Coppock and Stephanie Johnson, 2004.

National Research Council. *High-Performance Structural Fibers for Advanced Polymer Matrix Composites.* Committee on High-Performance Structural Fibers for Advanced Polymer Matrix Composites, 2005.

National Research Council. *The Limits of Organic Life in Planetary Systems.* Committee on the Limits of Organic Life in Planetary Systems. Committee on the Origins and Evolution of Life, 2007.

National Research Council of the National Academies. *Review of the NASA Astrobiology Institute.* Committee on the Review of the NASA Astrobiology Institute. Space Studies Board. Division on Engineering and Physical Sciences. Washington, D.C.: National Academies Press, 2007.

National Research Council. *The Scientific Context for Exploration of the Moon—Interim Report.* Committee on the Scientific Context for Exploration of the Moon. Space Studies Board. Division on Engineering and Physical Sciences. Washington, D.C.: National Academies Press, 2007.

Nealson, Kenneth H. and Radu Popa. "Introduction and Overview: What Do We Know for Sure?" *American Journal of Science* 305 (June, September, October, 2005): 449–66.

Nelson, David, and Michael Cox. *Lehninger Principles of Biochemistry* 3rd ed. New York: Worth Publishers, 2000.

Nicolaou, K. C. "Joys of Molecules." *Journal of Organic Chemistry* 70:18 (September, 2, 2005): 1225–58.

Nicolaou K. C., E. J. Sorensen, and N. Winssinger. "The Art and Science of Organic and Natural Products Synthesis." *Journal of Chemical Education* 75:10 (October 1998): PP.

Niele, Frank. *Energy—Engine of Evolution*. Amsterdam: Elsevier, 2005.

Nisbet, E. G., and C. M. R. Fowler. "The Early History of Life" *Treatise on Geochemistry*, Vol. 8. Edited by William H. Schlesinger. Amsterdam: Elsevier Publishing Company, 2003, 1–39.

Nisbet, E. G., and N. H. Sleep. "The Habitat and Nature of Early Life." *Nature* 409 (February 22, 2001): 1083–91.

Nobel Prize in Physics 2006. Information for the Public. The Royal Swedish Academy of Sciences. Accessed at http://nobelprize.org/nobel_prizes/physics/laureates/2006/info.html on February 8, 2008.

Nordhaus, William. "Critical Assumptions in the Stern Review on Climate Change." *Science* 317 (July 13, 2007): 201–2.

Norstog, Knut J., Ernest M. Gifford, and Dennis Wm. Stevenson. "Comparative Development of the Spermatozoids of Cycads and *Ginkgo biloba*." *Botanical Review* 70:1 (January–March 2004): 5–15.

Nye, Mary Jo. *Before Big Science: The Pursuit of Modern Chemistry and Physics, 1800–1940*. Cambridge, Mass: Harvard University Press, 1999.

O'Connor, Thomas Patrick. "Harold Morowitz Tackles Life—With the Help of a Few Friends." *Santa Fe Institute Bulletin* (Spring 2000): 2–5.

Oren, Aharon. "A Proposal for Further Integration of the Cyanobacteria under the Bacteriological Code." *Journal of Systemic and Evolutionary Biology* 54 (2004): 1895–902.

Ogg, James G. "Status of Divisions of the International Geologic Time Scale." *Lethaia* 37 (2004): 183–99.

Orgel, Leslie E. "Prebiotic Chemistry and the Origin of the RNA World." *Critical Reviews in Biochemistry and Molecular Biology* 39 (2004): 99–123.

———. "Self-Organizing Biochemical Cycles." *Proceedings of the National Academy of Sciences* 97:23 (November 7, 2000): 12503–07.

Orr, James C., Victoria J. Fabry, Olivier Aumont, Laurent Bopp, Scott C. Doney, Richard A. Feely, Anand Gnanadesikan, Nicolas Gruber, Akio Ishida, Fortunat Joos, Robert M. Key, Keith Lindsay, Ernst Maier-Reimer, Richard Matear, Patrick Monfray, Anne Mouchet, Raymond G. Najjar, Gian-Kasper Plattner, Keith B. Rodgers, and Christopher L. Sabine. "Anthropogenic

Ocean Acidification Over the Twenty-first Century and Its Impact on Calcifying Organisms." *Nature* 437, (Sept. 29, 2005): 681–86.

Pacala, Stephen, and Robert Socolow. "Stabilization Wedges: Solving the Climate Problem for the Next 50 Years with Current Technologies." *Science* 305 (August 13, 2004): 968–72.

Pagani, Mark, Ken Caldeira, David Archer, James C. Zachos. "An Ancient Carbon Mystery." *Science* 314 (December 8, 2006): 1556–57.

Palumbi, Stephen. "Humans as the World's Greatest Evolutionary Force." *Science* 293 (2001): 1786–90.

Papineau, Dominic, Jeffrey J. Walker, Stephen J. Mojzsis, and Norman R. Pace. "Composition and Structure of Microbial Communities from Stromatolites of Hamelin Pool in Shark Bay, Western Australia." *Applied and Environmental Microbiology* 71:8 (August 2005): 4822–32.

"Part-only sequence for BBa_I15010" http://parts.mit.edu/r/parts/partsdb/puttext .cgi. Accessed January 20, 2008.

Pauli, Wolfgang. "Remarks on the History of the Exclusion Principle." *Science* 103 (February 22, 1946): 213–15.

Pendleton, Yvonne J., and Dale P. Cruikshank. "Life from the Stars?" *Sky & Telescope* 87:3 (March 1994): 36–42.

Penisi, Elizabeth. "Synthetic Biology Remakes Small Genomes." *Science* 310 (November 4, 2005): 769–77.

Penzias, Arno A. "The Origin of the Elements." Nobel Lecture, December 8, 1978. In *Nobel Lectures, Physics 1971–1980,* ed. Stig Lundquist. Singapore: World Scientific Publishing Co., 1992: 444–57.

Petit, J. R., J. Jouzel, D. Raynaud, N. I. Barkov, J. M. Barnola, I. Basile, M. Bender, J. Chappellaz, M. Davis, G. Delaygue, M. Delmotte, V. M. Kotlyakov, M. Legrand, V. Y. Lipenkov, C. Lorius, L. Pépin, C. Ritz, E. Saltzman, and M. Stievenard. "Climate and Atmospheric History of the Past 420,000 Years From the Vostok Ice Core, Antarctica." *Nature* 399 (June 3, 1999): 429–36.

Pollack, Robert. *Signs of Life: The Language and Meanings of DNA.* New York: Houghton Mifflin Company, 1994.

Pollack, Robert. "The Emergence of Information in the Universe." Frontiers of Science lecture, presented at Columbia University, October 30, 2006.

Pollan, Michael. *The Botany of Desire: A Plant's-Eye View of the World.* New York: Random House, 2001.

———. *The Omnivore's Dilemma.* New York: Penguin Press, 2006.

Poole, Anthony. "My Name Is LUCA—The Last Universal Common Ancestor." ActionBioscience.org. September 2002. http://www.actionbioscience.org/ newfrontiers/poolepaper.html. Accessed December 11, 2007.

Principe, Lawrence. "Chemistry," In *History of Modern Science and Mathematics.* Edited by Brian S. Baigrie. New York: Charles Scribner's Sons, 2002.

"Profit-Proof Kevlar?" *Economist* (November 30, 1985): 100.

Pyne, Stephen J. *Fire: A Brief History.* Seattle: University of Washington Press, 2001.

Qian, Bin, Srivatsan Raman, Rhiju Das, Philip Bradley, Airlie J. McCoy, Randy J. Read, and David Baker, "High-Resolution Structure Prediction and the Crystallographic Phase Problem." *Nature* 450 (November 8, 2007): 259–64.

Rai, Arti, and James Boyle. "Synthetic Biology: Caught between Property Rights, the Public Domain, and the Commons. *PLoS Biology* 5:3 (March 2007): 0389–93.

Rank, D. M., C. H. Townes, and W. J. Welch. "Interstellar Molecules and Dense Clouds." *Science* 174 (December 10, 1971): 1083–101.

Rasmussen, Steen, Liaohai Chen, David Deamer, David C. Krakauer Norman H. Packard, Peter F. Stadler, and Mark A. Bedau, "Transitions from Nonliving to Living Matter." *Science* 303 (February 13, 2004) 963–65.

Raven, J. A., and D. Edwards. "Roots: Evolutionary Origins and Biogeochemical Significance." *Journal of Experimental Botany* 52 (March 2001): 381–401.

Raver, Anne. "Hardy Ginkgo Trees Are Fossils Minus the Rocks." *New York Times,* January 19, 1997.

Reisch, Marc S. "From Coal Tar to Crafting a Wealth of Diversity." *Chemical and Engineering News* 76:2 (January 12, 1998): 79.

Retallack, Gregory J. "Carbon Dioxide and Climate over the Past 300 Myr." *Philosophical Transactions of the Royal Society A* 260 (2002): 659–73.

Reyes, J. F., and M. A. Sepúlveda. "PM-10 Emissions and Power of a Diesel Engine Fueled with Crude and Refined Biodiesel from Salmon Oil." *Fuel* 85:12–13 (September 2006): 1714–19.

Rice, Richard. "Smokeless Powder: Scientific and Institutional Contexts at the End of the 19th Century." In Brenda J. Buchanan, *Gunpowder, Explosives and The State: A Technological History.* Burlington, Vt: Ashgate Publishing Co, 2006, 355–66.

Ridgwell, Andy J., Martin J. Kennedy, and Ken Caldeira. "Carbonate Deposition, Climate Stability, and Neoproterozoic Ice Ages." *Science* 302 (October 31, 2003): 859–62.

Ridgwell, Andy, and Richard E. Zeebe. "The Role of the Global Carbonate Cycle in the Regulation and Evolution of the Earth System." *Earth and Planetary Science Letters* 234 (2005): 299–315.

Ridley, Matt. *Genome: The Autobiography of a Species in 23 Chapters.* New York: Perennial, 1999.

Riebesell, Ulf, Ingrid Zondervan, Björn Rost, Philippe D. Tortell, Richard E. Zeebe, and François M. M. Morel. "Reduced Calcification of Marine Plankton in

Response to Increased Atmospheric CO_2." *Nature* 407 (September 21, 2000): 364–67.

Robinson, Richard. "A Smart Mutation Scheme Produces Hundreds of Functional Proteins." *PLoS Biology* 4:5 (May 2006): 0663.

Rode, A.V., E. G. Gamaly, A. G. Christy, J. Fitz Gerald, S. T. Hyde, R. G. Elliman, B. Luther-Davies, A. I. Veinger, J. Androulakis, and J. Giapintzakis. "Strong Paramagnetism and Possible Ferromagnetism in Pure Carbon Nanofoam Produced by Laser Ablation." *Journal of Magnetism and Magnetic Materials* 290–91:1 (April 2005): 298–301.

Romm, Joseph. *Hell and High Water: Global Warming—The Solution and the Politics—and What We Should Do.* New York: William Morrow, 2007.

Ronen, Avraham. "Domestic Fire as Evidence for Language." In *Neanderthals and Modern Humans in Western Asia,* edited by Akazawa Takeru, Kenichi Aoki, and Ofer Bar-Yosef, New York: Plenum Press, 1998, 439–47.

Rösler, Wolfram. "The Hello World Collection." http://www.roesler-ac.de/wolfram/hello.htm. Accessed January 20, 2008.

Rost, B., and U. Riebesell. "Coccolithophores and the Biological Pump: Responses to Environmental changes." In *Coccolithophores: From Molecular Processes to Global Impact,* edited by Hans R. Thierstein and Jeremy R. Young. Berlin: Springer, 2003, 99–125.

Roth, Katherine C., David M. Meyer, and Isabel Hawkins. "Interstellar Cyanogen and the Temperature of the Cosmic Microwave Background Radiation." *Astrophysical Journal* 413 (August 20, 1993): L67–L71.

Rothman, Daniel H. "Atmospheric Carbon Dioxide Levels for the Last 500 Million Years." *Proceedings of the National Academy of Sciences* 99:7 (April 2, 2002): 4167–417.

———. "Global Biodiversity and the Ancient Carbon Cycle." *Proceedings of the National Academy of Sciences* 98:8 (April 10, 2001): 4305–10.

Royer, Dana L. "CO_2-Forced Climate Thresholds during the Phanerozoic." *Geochimica et Cosmochimica Acta* 70 (2006): 5665–75.

Royer, Dana L., Robert A. Berner, and Jeffrey Park. "Climate Sensitivity Constrained by CO_2 Concentrations over the Past 420 Million Years." *Nature* 446 (March 29, 2007): 2310–13.

Royer D. L., S. L. Wing, D. J. Beerling, D. W. Jolley, P. L. Koch, L. H. Hickey, and R. A. Berner "Paleobotanical Evidence for Near Present Day Levels of Atmospheric CO_2 During Part of the Tertiary." *Science* 292 (June 22, 2001): 2310–13.

Royer, Dana, L., Leo J. Hickey, and Scott Wing. "Ecological Conservatism in the 'Living Fossil' Ginkgo." *Paleobiology* 29:1 (2003): 84–104.

Ruddiman, William F. *Plows, Plagues, and Petroleum: How Humans Took Control of Climate.* Princeton,: Princeton University Press, 2005.

Rutan, Burt. "Burt Rutan: Entrepreneurs are the Future of Space Flight." TED. Monterey (February 2006). http://www.ted.com/index.php/talks/view/id/4. Accessed May 23, 2007.

Ryan, Frank. *Darwin's Blind Spot: Evolution Beyond Natural Selection.* New York: Houghton Mifflin Company, 2002.

Sagan, Carl. *Cosmos.* New York: Ballantine Books, 1980.

Salpeter, Edwin E. "A Generalist Looks Back." *Annual Review of Astronomy and Astrophysics* 40 (2002): 1–25.

———. "Energy Production in Stars." *Annual Review of Nuclear Science* 2 (December 1953): 41–65.

Schidlowski, Manfred. "Carbon Isotopes as Biogeochemical Recorders of Life over 3.8 Ga of Earth History: Evolution of a Concept." *Precambrian Research* 106 (2001): 117–34.

Schmidt-Bleek, Friedrich Bio. *The Fossil Makers.* Translated by Reuben Deumling. Birkhäuser, 1993. http://www.factor10-institute.org/seitenges/Pdf-Files.htm. Accessed February 8, 2008.

Schmitt, Harrison H. "Mining the Moon." *Popular Mechanics* (October 2004): 56–61.

Schneider, Eric D., and Dorion Sagan. *Into the Cool: Energy Flow, Thermodynamics and Life.* Chicago: University of Chicago Press, 2005.

Schneider, Stephen H. "The Greenhouse Effect: Science and Policy." *Science* 243 (February 10, 1989): 771–81.

Schopf, J. William, ed. *Life's Origin: The Beginnings of Biological Evolution.* Los Angeles: University of California Press, 2002.

Schopf, J. William "Fossil Evidence of Archaean Life." *Philosophical Transactions of the Royal Society A* 361 (June 29, 2006): 869–85.

———. "Solution to Darwin's Dilemma: Discovery of the Missing Precambrian Record of Life." *Proceedings of the National Academy of Sciences* 97:13 (June 20, 2000): 6947–53.

Sebald, W. G. *The Rings of Saturn.* Translated by Michael Hulse. New York: New Directions, 1999.

Second International Conference on Synthetic Biology. May 20–22, 2006. University of California–Berkeley. Webcast.berkeley. http://webcast.berkeley.edu/event_details.php?webcastid=15766. Accessed November 13, 2006.

Seeds, Michael A. *Foundations of Astronomy* 8th ed. Belment, Calif.: Thomson Brooks/Cole, 2004.

Self, Sydney B. "Researchers Fomenting Another 'Revolution' in Oil Refining Process." *Wall Street Journal*, March 4, 1940, 1.

Sephton, Mark A. "Organic Compounds in Carbonaceous Meteorites." *Natural Product Reports* 19 (2002): 292–311.

Service, Robert F. "The Race for the $1000 Genome." *Science* 311 (March 17, 2006): 1544–46.

Shatto, Rahilla C. A., and Niall C. Slowey. "Thinking Big: Coccolithophores May Be Small but They Know How to Get Attention." *Quarterdeck* 5:2 (1997), http://www-ocean.tamu.edu/Quarterdeck/QD5.2/shatto-slowey.html. Accessed February 8, 2008.

Shallenberger, Robert S. *Taste Chemistry.* Glasgow: Blackie Academic and Professional, 1993.

Shreeve, James. "The Greatest Journey." *National Geographic* (March 2006): 61–69.

Shreeve, James. "Reading Secrets of the Blood." *National Geographic* 209:3 (March 2006): 70–73.

Siesser, William G. "Historical Background of Coccolithophore Studies." In *Coccolithophores*, edited by Amos Winter, and William Siesser. Cambridge: Cambridge University Press, 1994, 1–11.

Simmons, Matthew. *Twilight in the Desert.* Hoboken, N.J.: John Wiley & Sons, 2005.

Simonson, Anne B., Jacqueline A. Servin, Ryan G. Skophammer, Craig W. Herbold, Maria C. Rivera, and James A. Lake. "Decoding the Genomic Tree of Life." *Proceedings of the National Academy of Sciences* 102:1 (May 3, 2005): 6608–13.

Singh, Simon. *Big Bang: The Origin of the Universe.* New York: Fourth Estate, 2004.

Smil, Vaclav. "Energy at the Crossroads." Background notes for a presentation at the Global Science Forum Conference on Scientific Challenges for Energy Research, Paris, May 17–18, 2006.

———. *Cycles of Life: Civilization and the Biosphere.* New York: Scientific American Library, 2001.

———. *Energy: A Beginner's Guide.* Oxford: One World Publications, 2006.

———. *Energy at the Crossroads.* Cambridge, Mass.: MIT Press, 2005.

Smith, Eric, and Harold J. Morowitz. "Universality in Intermediary Metabolism." *Proceedings of the National Academy of Sciences* 101:36 (September 7, 2004): 13169–73.

Smith, Hamilton O., Robert Friedman, and J. Craig Venter. "Biological Solutions to Renewable Energy." *Bridge Archives* 33:2 (Summer 2003). Accessed from http://www.nae.edu/nae/bridgecom.nsf/weblinks/MKUF-5NTMX9?OpenDocument on May 22, 2007.

Smith, John Maynard, and Eörs Szathmáry. *The Origins of Life: From the Birth of Life to the Origin of Language.* New York: Oxford University Press, 2000.

Smith, J. V., and Y. Luo. "Studies on Molecular Mechanisms of *Ginkgo biloba* Extract." *Applied Microbiological Biotechnology* 64 (2004): 465–72.

Snyder, Lewis E. "Interferometric Observations of Large and Biologically Interesting Interstellar and Cometary Molecules." *Proceedings of the National Academy of Sciences* 103:33 (August 15, 2006): 12243–48.

Snyder, L. E., F. J. Lovas, J. M. Hollis, D. N. Friedel, P. R. Jewell, A. Remijan, V. V. Ilyushin, E. A. Alekseev, and S. F. Dyubko. "A Rigorous Attempt to Verify Interstellar Glycine." *The Astrophysical Journal*, 619:2 (Feb. 1, 2005): 914–30.

Sommers, Michael G., Michael E. Dollhopf, and Susanne Douglas. "Freshwater Ferromanganese Stromatolites from Lake Vermilion, Minnesota: Microbial Culturing and Environmental Scanning Electron Microscopy Investigations." *Geomicrobiology Journal* 19:4 (July–August 2002): 407–27.

Spear, Ray. "The Most Important Experiment Ever Performed by an Australian Physicist." *Physicist* 39:2 (March–April 2002): 35–41.

Spitz, Peter H. *Petrochemicals: The Rise of an Industry.* New York: John Wiley & Sons, 1988.

Spolyar, Douglas, Katherine Freese, and Paolo Gondolo. "Dark Matter and the First Stars: A New Phase of Stellar Evolution." *Physical Review Letters* 100 (February 8, 2008): 141.

Srinivasan, V., and H. Morowitz. "Ancient Genes in Contemporary Persistent Microbial Pathogens." *Biological Bulletin* 210 (February 2006): 1–9.

Stanley, Dick. "Nobel Just the First Product of Buckminsterfullerene." *Austin American-Statesman*, (October 10, 1996): A1.

Stern Review on the Economics of Climate Change. http://www.sternreview.org .uk/. Accessed March 2007.

Stern, Nicholas, and Chris Taylor. "Climate Change: Risk, Ethics, and the Stern Review." *Science* 317 (July 13, 2007): 203–4.

Stoneley, R. "A Review of Petroleum Source Rocks in Parts of the Middle East." Edited by J. Brooks and A. J. Fleet. *Marine Petroleum Source Rocks*, Geological Society Special Publication No 26. Oxford: Blackwell Science Publications, 1987.

Summons, Roger E., Linda L. Jahnke, Janet M. Hope, and Graham A. Logan. "2-Methylhopanoids as Biomarkers for Cyanobacterial Oxygenic Photosynthesis." *Nature* 400 (August 4, 1999): 554–57.

Szathmáry, Eörs. "In Search of the Simplest Cell." *Nature* 433 (February 3, 2005): 469–70.

Szostak, Jack W., David P. Bartel, and P. Luigi Luisi. "Synthesizing Life." *Nature* 409 (January 18, 2001): 387–90.

Tanford, Charles, and Jacqueline Reynolds. *Nature's Robots: A History of Proteins.* Oxford: Oxford University Press, 2001.

Tanner, David, James A. Fitzgerald, and Brian R. Phillips. "The Kevlar Story—An

Advanced Materials Case Study." *Angewandte Chemie International Edition English Advanced Materials* 28:5 (1989): 649–54.

Taylor, Richard C., and V. J. Rowntree. "Temperature Regulation and Heat Balance in Running Cheetahs: A Strategy for Sprinters?" *American Journal of Physiology* 224:4 (April 1973) 848–51.

Tennant, Smithson. "On the Nature of the Diamond." *Philosophical Transactions of the Royal Society of London* 87 (1797): 123–27.

Tierney, Christine. "Ford Slams Toyota on Hybrids." *Detroit News,* August 8, 2005. http://www.detnews.com/2005/autosinsider/0508/08/A01-272872.htm. Accessed January 7, 2008.

Thomas, Patricia. "The Chemical Biologists." *Harvard Magazine* March–April 2005. Accessed from http://harvardmagazine.com/2005/03-pdfs/0305-38 .pdf on September 26, 2006.

Todar, Kenneth. "The Bacterial Flora of Humans." *Todar's Online Textbook of Bacteriology* (2007): http://textbookofbacteriology.net/normalflora.html. Accessed February 8, 2008.

Todd, O. M., and Sir John Cornforth. "Robert Burns Woodward." *Biographical Memoirs of Fellows of the Royal Society,* Vol. 27. (November 1981): 628–95.

Tol, Richard. "The Marginal Damage Costs of Carbon Dioxide Emissions: An Assessment of the Uncertainties." *Energy Policy* 33 (2005): 2064–74.

Tomitani, Akiko, Andrew H. Knoll, Colleen M. Cavanaugh, and Terufumi Ohno. "The Evolutionary Diversification of Cyanobacteria: Molecular-Phylogenetic and Paleontological Perspectives." *Proceedings of the National Academy of Sciences* 103:14 (April 4, 2006) 5442–47

Toone, Eric J., ed. *Protein Evolution.* Hoboken, N. J.: John Wiley & Sons Inc., 2007.

Townes, Charles H. "A Physicist Courts Astronomy." *Annual Review of Astronomy and Astrophysics* 35 (1997): xiii–xliv.

Trainer, Melissa G., Alexander A. Pavlov, H. Langley DeWitt, Jose L. Jimenez, Christopher P. McKay, Owen B. Toon, and Margaret A. Tolbert. "Organic Haze on Titan and the Early Earth." *Proceedings of the National Academy of Sciences* 103:48 (November 28, 2006): 18035–42.

Tralau, Hans. "Evolutionary Trends in the Genus *Ginkgo.*" *Lethaia* 1:1 (January 1968): 63–101.

Travers, M. J., M. C. McCarthy, P. Kalmus, C. A. Gottlieb, and P. Thaddeus. "Laboratory Detection of the Cyanopolyyne $HC_{13}N$." *Astrophysical Journal Letters* 472 (November 1996): L61–L62.

Tumpey, Terrence M., Christopher F. Basler, Patricia V. Aguilar, Hui Zeng, Alicia Solórzano, David E. Swayne, Nancy J. Cox, Jacqueline M. Katz, Jeffery K. Taubenberger, Peter Palese, Adolfo García-Sastre. "Characterization of the

Reconstructed 1918 Spanish Influenza Pandemic Virus." *Science* 310 (October 7, 2005): 77–80.

Tweney, Ryan D., and David Gooding eds. *Michael Faraday's "Chemical Notes, Hints, Suggestions and Objects of Pursuit" of 1822.* London: Peter Peregrinus, 1991.

Tyrrell, Toby, and Richard E. Zeebe. "History of Carbonate Ion Concentration over the Last 100 Million Years." *Geochimica et Cosmochimica Acta* 68:17 (2004): 3521–30.

Tyrrell, Toby, and Jeremy Young. "Coccolithophores." (Forthcoming.)

Tyson, Neil de Grasse, and Donald Goldsmith. *Origins: 14 Billion Years of Cosmic Evolution.* New York: Norton, 2005.

U.S. Department of Energy. *Basic Research Needs for Clean and Efficient Combustion of 21st Century Transportation Fuels.* Report of the Basic Energy Sciences Workshop on Basic Research Needs for Clean and Efficient Combustion of 21st Century Transportation Fuels. Accessed from http://www.sc.doe.gov/bes/reports/files/CTF_rpt.pdf on June 4, 2007.

United States Geological Survey. "Descriptions and Origins of Selected Principal Building Stones of Washington." http://pubs.usgs.gov/gip/stones/ descriptions .html. Accessed October 31, 2007.

Valley, John W. "A Cool Early Earth?" *Scientific American* 293:4 (October 2005): 58–65.

van Beek, Teris A., ed. *Ginkgo Biloba.* Amsterdam: Harwood Academic Publishers, 2000.

Varchaver, Nicholas. "Chemical Reaction." *Fortune* April 2, 2007, 52–58. Accessed from http://money.cnn.com/magazines/fortune/fortune_archive/2007/04/ 02/8403424/index.htm on April 23, 2007.

Veizer, Jan, Yves Godderis, and Louis M. Francois. "Evidence for Decoupling of Atmospheric CO_2 and Global Climate During the Phanerozoic Eon." *Nature* 408 (December 7, 2000): 698–701.

Venables, Michelle C., Juul Achten, and Asker E. Jeukendrup. "Determinants of Fat Oxidation During Exercise in Healthy Men and Women: A Cross-Sectional Study." *Journal of Applied Physiology* 98 (2005): 160–67.

Venter, J. Craig, Karin Remington, John F. Heidelberg, Aaron L. Halpern, Doug Rusch, Jonathan A. Eisen, Dongying Wu, Ian Paulsen, Karen E. Nelson, William Nelson, Derrick E. Fouts, Samuel Levy, Anthony H. Knap, Michael W. Lomas, Ken Nealson, Owen White, Jeremy Peterson, Jeff Hoffman, Rachel Parsons, Holly Baden-Tillson, Cynthia Pfannkoch, Yu-Hui Rogers, Hamilton O. Smith. "Environmental Genome Shotgun Sequencing of the Sargasso Sea." *Science* 304 (April 2, 2004): 66–74.

Vernadsky, Vladimir Ivanovich. "A Few Words about the Noösphere." In *Scientific Thought as a Planetary Phenomenon*, edited by A. L. Yashin. Moscow: Nauka, 1991. Web version "From the Archive of V. I. Vernadsky," Author translation. http://vernadsky.lib.ru/e-texts/archive/noos.html. Accessed May 26, 2007.

Wächtershäuser, Gunter. "Evolution of the First Metabolic Cycles." *Proceedings of the National Academy of Sciences* 87 (January 1990): 200–204.

Waldrop, M. Mitchell. *Complexity: The Emerging Science at the Edge of Order and Chaos*. New York: Touchstone, 1992.

Waldrop, M. Mitchell. "Did Life Really Start Out in an RNA World?" *Science* 246 (December 8, 1989): 1248–49.

———. "How Do You Read from the Palimpsest of Life?" *Science* 246 (November 3, 1989): 578–79.

Wallerstein, George, Icko Iben Jr., Peter Parker, Ann Merchant Boesgaard, Gerald M. Hale, Arthur E. Champagne, Charles A. Barnes, Franz Käppeler, Verne V. Smith, Robert D. Hoffman, Frank X. Timmes, Chris Sneden, Richard N. Boyd, Bradley S. Meyer, David L. Lambert. "Synthesis of the Elements in Stars: Forty Years of Progress." *Reviews of Modern Physics* 69:4 (October 1997): 995–1084.

Walling, Olivia. "Research at the Kellogg Radiation Laboratory, 1920s–1960s: A Small Narrative of Physics in the Twentieth Century." Ph.D. diss. University of Minnesota, 2005.

Walter, Chip. *Thumbs, Toes, and Tears and Other Traits That Make Us Human*. New York: Walker & Co., 2006.

Walter, Katie. "The Internal Combustion Engine at Work: Modeling Considers All Factors." *Science and Technology Review* (Lawrence Livermore National Laboratory) (December 1999): 4–10.

Ward, Peter. *Out of Thin Air*. Washington, D.C.: Joseph Henry Press, 2006.

Weart, Spencer. *The Discovery of Global Warming*. Spencer Weart & American Institute of Physics, 2003–2006. http://www.aip.org/history/climate. Accessed April 22, 2006.

Weil, Andrew. *Eating Well for Optimum Health: The Essential Guide to Bringing Health and Pleasure Back to Eating*. New York: Quill, 2001.

Weinstock, George M. "ENCODE: More Genetic Empowerment." *Genome Research* 17 (2007): 667–68.

Weitzman, Martin. "The Stern Review of the Economics of Climate Change." *Journal of Economic Literature* 4:3 (September 2007). http://www.economics.harvard.edu/faculty/weitzman/files/JELSternReport.pdf. Accessed December 8, 2007.

Wells, Spencer. "Out of Africa." *Vanity Fair* 563 (July 2007): 110–14.

Westbroek, Pieter. *Life as a Geological Force.* New York: Norton, 1991.

Whitesides, George M. and Bartosz Grzybowski. "Self-Assembly at All Scales." *Science* 295 (March 29, 2002): 2418–21.

Whitman, William B., David C. Coleman, and William J. Wiele. "Prokaryotes: The Unseen Majority." *Proceedings of the National Academy of Sciences* 95 (1998): 6578–83.

Wilde, Simon A., John W. Valley, William H. Peck, and Colin M. Graham. "Evidence From Detrital Zircons for the Existence of Continental Crust and Oceans on the Earth 4.4 Gyr Ago." *Nature* 409 (January 11, 2001): 175–78.

Wilford, John Noble. "Ancient Tree Yields Secrets of Potent Healing Substance." *New York Times,* March 1, 1988, C3.

Wills, Christopher, and Jeffrey Bada. *The Spark of Life: Darwin and the Primeval Soup.* Cambridge, Mass.: Perseus Publishing, 2000.

Wilson, David S., and Jack W. Szostak. "In Vitro Selection of Functional Nucleic Acids." *Annual Review of Biochemistry* 68 (1999): 611–47.

Wilson, Edwin O. *Consilience: The Unity of Knowledge.* New York: Vintage, 1998.

———. *The Future of Life.* New York: Knopf, 2002.

Wilson, Rachel I., and Roger A. Nicoli. "Endocannabinoid Signaling in the Brain." *Science* 296 (April 26, 2002): 678–82.

Wilson, Rebecca M., and Samuel J. Danishefsky. "Small Molecule Natural Products in the Discovery of Therapeutic Agents: The Synthesis Connection." *Journal of Organic Chemistry* 71:22 (October 27, 2006): 8329–51.

Wilson, R. W., K. B. Jefferts, and A. A. Penzias. "Carbon Monoxide in the Orion Nebula." *Astrophysical Journal* 161 (July 1970): L43–44.

Wilson, S. S. "Bicycle Technology." *Scientific American* 228 (March 1973): 81–91.

Woese, Carl. *The Genetic Code: The Molecular Basis for Genetic Expression.* New York: Harper and Row, 1967.

Woese, C. R., O. Kandler, and M. L. Wheelis. "Towards a Natural System of Organisms: Proposal for the Domains Archaea, Bacteria, and Eucarya." *Proceedings of the National Academy of Sciences* 87:12 (June 1990): 4576–79.

Wolf, Yuri I., Igor B. Rogozin, Nick V. Grishin, and Eugene V. Koonin. "Genome Trees and the Tree of Life." *Trends in Genetics,* 18:9 (September 2002): 472–79.

Womack, James P., Daniel T. Jones, Daniel Roos, and Donna Sammons Carpenter. *The Machine That Changed the World.* New York: Rawson Associates, Macmillan Publishing Company, 1990.

Woodhead, James A., ed. *Geology.* Pasadena, Calif.: Salem Press, 1999.

Woodman, A. G. "The Exact Estimation of Atmospheric Carbon Dioxide: A Survey." *Technology Quarterly and Proceedings of the Society of Arts* 17:1 (March 1904): 258–69.

Woodward, Robert Burns. "The Arthur C. Cope Award Lecture," In Otto Theodore Benfey and Peter J. T. Morris, *Robert Burns Woodward: Architect and Artist in the World of Molecules.* Philadelphia: Chemical Heritage Foundation, 2001, 418–39.

Yergin, Daniel. *The Prize: The Epic Quest for Oil, Money and Power.* New York: Touchstone, 1992.

Young, Jeremy R. "Crystal Assembly and Phylogenetic Evolution in Heterococcoliths." *Nature* 356 (April 9, 1992): 516–18.

Young, Jeremy R., and Karen Henriksen. "Biomineralization Within Vesicles: The Calcite of Coccoliths." *Biomineralization* 54 (2003): 189–215.

Zahnle, K., M. Claire, and D. Catling. "Methane, Sulfur, Oxygen, Ice: The Loss of Mass-Independent Fractionation in Sulfur Due to a Palaeoproterozoic Collapse of Atmospheric Methane." *Geobiology* (2006).

Zeebe, Richard E., and Pieter Westbroek. "A Simple Model for the $CaCO_3$ Saturation State of the Ocean: The 'Strangelove,' the 'Neritan,' and the 'Cretan' Ocean." *G3: Geochemistry Geophysics Geosystems* 4:12 (December 12, 2003): 1–26.

Zhang, Mao–Xi, Philip E. Eaton, and Richard Gilardi. "Hepta- and Octanitrocubanes." *Angewandte Chemie International Edition in English* 39:2 (2000): 401–4.

Zhou, Zhiyan, and Shaolin Zheng. "The Missing Link in Ginkgo Evolution." *Nature* 423, (June 19, 2003): 821–22.

Zimmer, Carl. *Evolution: Triumph of an Idea.* New York: Harper Perennial, 2001.

Zimmer, Carl. "What Came Before DNA?" *Discover* (June 26, 2004): Accessed from http://discovermagazine.com/2004/jun/cover/article-print on June 12, 2007.

Interviews

Formal or informal, by phone, in person, over e-mail, or by post, communication with the following people proved critical to researching *The Carbon Age*:

Lihini Aluwihare

Bob Arnett

Frances Arnold

Gustaf Arrhenius

Kenneth Arrow

Steven Aldrich

Norman Augustine

Jesse Ausubel

Farooq Azam

Jeffrey Bada

Phil Baran
Luann Becker
Herb Belin
Ruth Greenspan Bell
Dean vanden Biesen
Ulrika Bjorksten
Mary Ellen Bowden
Jochen Brocks
Wallace Broecker
Robert Berner
Geoffrey Burbidge
Margaret Burbidge
Ken Caldeira
William Carroll
Jon Clardy
Jim Cleaves
David Cole
James Collins (Boston University)
James Collins (National Science
    Foundation)
E. J. Corey
Elizabeth Cottrelle
Michael Crimmins
David Crisp
Robert Curl
Phil DeCola
Peter Del Tredici
Gabriel Eichler
Drew Endy
Myer Ezrin
Frank Ferrari
James Ferris
William Friedman
Hans Fynbo
Vlodek Gabara
Newt Gingrich
Susan Golden
John Gowdy
Bob Gower

Harry Gray
Woody Hastings
John Hatleberg
John Hayes
Robert Hazen
David Helfand
Russell Hemley
John Hendricks
Eric Herbst
Dudley Herschbach
Roald Hoffmann
Susan Hovorka
John Hunt
Shirley Ann Jackson
Filip Johnsson
Gerald Joyce
Erland Källén
James Kasting
Ralph Keeling
Charles Kennel
David Kirsch
Joseph Kirschvink
Johan Kleman
Andrew Knoll
Ray Kopp
Harry Kroto
Karen Kwitter
James Leighton
Allen Lim
John Lough
Cesare Marchetti
Harold Morowitz
Gregory McRae
Joel Mokyr
Koji Nakanishi
Richard Norris
Christine O'Brien
Mark Ohman
Leslie Orgel

Joseph Patterson
Yvonne Pendleton
Billy Pizer
Anthony Poole
Gregory Retallack
David Rind
Alan Rocke
Joseph Romm
Gunhild Rosqvist
Dana Royer
Stuart Schreiber
Edwin Salpeter
Jeff Severinghaus
James Shigley
Lester Shubin
Richard Smalley
Kenneth Smith
David Steenkamer

Roger Summons
Jack Szostak
Lewis Snyder
Vaclav Smil
Gwynedd A. Thomas
Mark Thiemens
Toby Tyrrell
Lori Wagner
Olivia Walling
Ray Weiss
Paul Wennberg
George M. Whitesides
George T. Whitesides
Scott Wing
George Woodwell
Jeremy Young
Yuk Yung

INDEX

A Note on the Author

Eric Roston is a journalist and science writer. As a *Time* magazine reporter, he covered some of the most high profile stories of recent times, including the attacks of 9/11 and the 2004 presidential campaign. A member of the National Association of Science Writers (NASW) and the Society of Environmental Journalists (SEJ), he regularly covered energy, climate, technology, science, and business for the magazine before leaving to write this book. Roston graduated from Columbia University, with a major in the history of the World Wars, and earned a Master's degree in Russian literature and linguistics. Born and raised outside of Chicago, he lives in Washington, D.C., with his wife, Karen Yourish, and their daughter.